LET ME

EAT

CAKE

PAUL ARNOTT

LET ME

EAT

CAKE

with all best wishes

Paul Arnott

2007

S

SCEPTRE

First published in Great Britain in 2007 by Hodder and Stoughton
A division of Hodder Headline

A Sceptre Book

A CIP catalogue record for this title is available from the British Library

ISBN 0 340 92397 0
9 780340 923979

Typeset in Sabon by Hewer Text UK Ltd, Edinburgh
Printed and bound by Clays Ltd, St Ives plc

Hodder Headline's policy is to use papers that are natural, renewable
and recyclable products and made from wood grown in sustainable
forests. The logging and manufacturing processes are expected to
conform to the environmental regulations of the country of origin

Hodder and Stoughton Ltd
A division of Hodder Headline
338 Euston Road
London NW1 3BH

To Jake, Benjamin, Sasha and Tara

Silvertown

In the kitchen that my mother ran with all the pride of a *chef cuisiniers* at Versailles, even if it was in white-bread Bromley, she provided the plenty she had gone without herself. Her childhood had been marred by chocolate factories converted to make tyres for army trucks with the knowledge that the Yanks on Piccadilly seemed to have as much chocolate as they wanted, even if it was a Hershey bar and tasted like soap.

Ten years after the war my mother, a secretary, and my father, a motor insurance manager, adopted their first son, my brother. Six years later they came home from the shop with me. They knew that raising adopted children would present many challenges but they would underpin the project by giving us the best they could afford. Her larder would always be full and she wouldn't stint on the sugar.

My job was to eat what was put in front of me. I was the child of relatively old parents with deep roots in another age. My mother's name was Betty. Her father was one of thirteen from Bermondsey. They had names like Arthur, Queenie, Vi, Fred, Lil, Doll and so on. When they came to tea on a Sunday afternoon a spread was laid out akin to a suburban feast in the year of Queen Victoria's Golden Jubilee.

The wobbling centrepiece was an aspic jelly with egg and tinned salmon – always John West – set in a mould. When the tin was heated in lukewarm water a discreet salmon fart puffed across the kitchen and Betty turned it on to a plate ready for the serving hatch. Following it through were beetroot, cucumber and lettuce. Sliced and buttered white bread. Pilchards in tomato sauce. Salads were never mixed, but there was vinegar or – and I loved it – salad cream. It did not occur to me then that it was full of sugar. To a child it made rabbit food taste divine.

The dining room was dark but the French windows were thrown open to the garden beyond and the extended family tucked in. As they ate, they watched the frogs jump from pad to pad in the green-tinged pond, the odd goldfish making a play for a fly. I was self-appointed wasp-killer-in-chief, a small game hunter, deadly with a flick of a tea towel as I ended the lives of those selfish stingers set on diving headfirst into the cut-glass dish of salad cream.

I can barely recall the conversations over that table, except that my father always seemed to be making people laugh. I do remember once that Uncle Fred (a Stalinist who took the *Morning Star* during the week) had read an article about bowel cancer in his *Sunday Express* – an early example of health journalism ruining a good meal – and telling us about his movements as we ate. Betty was sincerely disgusted and never entirely forgave him.

I ate well but not greedily. Greed was held in reserve. I knew what remained in the larder. Those afternoons hosting the major family members were like feast days in more observant families, and I had been in attendance in the kitchen all morning. The sweet counterpart to the aspic jelly was to be a huge sherry trifle. It would be the occasion of terrible disappointment to my mother if one of our guests did not shriek, 'Oh, Betty, how much sherry have you put in this?' followed by a roar of laughter and elderly ladies saying, 'You'll have me tiddly, you will.'

I could have told them how much. Gallons. As much as the

sponge cake lining the base of the glass dish could absorb without disintegrating. Then pints of jelly, filled with tinned mandarin oranges, a thick layer of cold custard, cream, and some flaked almonds.

But this was just the beginning. All morning, meringues had been cooled from the oven, and wedged together with cream. Malt loaf had been sliced and loaded with thick English butter. At the crucible of the kitchen there was a wonder device, the Kenwood Chef mixer. A bowl would be filled with flour and sugar and butter and more sugar, dark brown now, and sultanas and raisins and, if I was really lucky, glacé cherries.

My reward for what little help I had been – keeping out of the way chasing the dog round the garden or bashing a football against a wall – was the paddle. This was the silver churning part screwed into the Kenwood which, on the recommendation of Fanny Craddock, was easing the strain of the brawny-armed housewife. This paddle was mine and I flicked it with a delicate tongue, not leaving a trace of cake mix behind.

This teatime banquet was what you got when you came to the Arnott house for the day, and I was proud of it. Nobody went home hungry. Indeed, most must have lapsed into indolence when they got back to Eltham.

Like contented children everywhere, though, I took it all for granted. It would be years before I met the latchkey kids, heard of convenience meals, or would even consider the social horror of eating alone. Perhaps there was more sugar in all this than might be considered ideal by modern standards, but that is not why I am, ahem, beyond my 'target weight' today. If there was a fault in what Betty did for us it was that it made life beyond childhood terribly complicated. I would never eat as well again, and without the services of a full-time cook – which Betty had been – it doesn't seem possible that I ever could. Mrs Bridges from *Upstairs Downstairs* did some of my mother's work, but without the washing and cleaning and shopping that went with it. When women everywhere else were sensing liberation, in my house one

3

person was working harder than an Edwardian maid to keep three men living like kings. It has taken me until the next century to understand that.

At the end of one of those long hot afternoons of plenty my father did something which I now realise good husbands should do more often. He knew that Betty had been slaving since dawn and, though she took great satisfaction from this, no woman wants the perfect day to end with her alone at the washing up.

'We'll go up to Greenwich,' he said. 'Take old George for a spin.'

George Barker lived in an alms house down on the river Thames near Deptford. The sun was still well above the horizon as we collected him. He walked like a Meccano man because he had lost both legs in the war. He could only fit his artificial limbs into the front seat, and with the dog in the back with my mother and brother it was reasoned that I should sit on his lap.

That ten-minute journey down the Creek Road was the most uncomfortable of my life. I could feel every strut and fret of his wooden legs through my young buttocks. We sat on the escarpment by General Wolfe's statue looking west into the sunset, the wreck of the Isle of Dogs across the river, a mess of dead warehouses and scrub which would not see Canary Wharf for another twenty-five years. This was how I first smelled the Sunday pong of Greenwich, which every resident can describe, that sometimes drifts south well beyond Blackheath.

Used only to the safe aromas of Bromley I asked what this industrial whiff was. Somehow it was catching my imagination. It was deeply pungent, sweet, treacly, and heavy in the hot air. George Barker sat down on a bench next to me, his sharp knees skinny in a well-creased suit, which looked like his demob pinstripe from thirty years before.

'That's the Tate & Lyle factory in Silvertown, Paul. Down over there. See that church. Just follow the line down to the riverside, and there it is, like a little country. Every Sunday they open up the vats and give them a good clean. That's what you can

4

smell. You should try it down in Deptford sometime – still hanging about come the Monday market.'

I had heard the word Silvertown, and started thinking about cowboys and Indians and runaway trains in Colorado. I felt odd on this ledge, looking down over London, as if it were falling away from me.

How could I have understood that in that same factory the vats which were being cleansed would tomorrow be inverting sucrose solution again, splitting glucose molecules through great heat, and pouring the resultant partially inverted sugar syrup into the green and gold tins of Lyle's Golden Syrup. My first love, my morphia, my obsession, my destiny.

Agonistes

When it was first sold in the late nineteenth century Golden Syrup came in wooden casks and was available only to Lyle's factory workers and to the East-enders who lived nearby. Like many other renowned sweet products (Nutella and Flake for example) it was never meant to exist at all; it was a by-product.

The ferociously ambitious Abram Lyle had sent his five sons from Glasgow to London to build a sugar refinery, and they discovered that in the process of refining sugar cane from the West Indies a treacly deposit would congeal at the bottom of their vats. At first they were glad to be able to get rid of it to sweet-toothed locals, but it soon became a desirable and lucrative product in its own right.

The tin designed in 1880 to carry the syrup beyond East London and across the Empire is one of the longest surviving works of graphic art in manufacturing history. Green with gold adornment, it is lent gravitas by the image of the rotting lion set in an oval frame beneath the shimmering white and gold Lyle's Golden Syrup logo.

This image was inspired by the story of Samson in the Book of Judges who, travelling to Timnah to get engaged to a Philistine, was attacked by a lion in a vineyard. Filled with a surge of power from the Lord, Samson tore the beast apart with his bare hands. 'As if it were a young goat', the Bible claims, which still seems quite a lot harder than a phone directory.

On the way to his wedding a few days later Samson went to gloat over the dead lion and found that a swarm of bees had made a hive in its flesh and produced honey. He scooped this out with his fingers and ate it. Boastfully, he set a riddle inspired by this for his bolshy Philistine in-laws at his wedding do. 'Out of the eater came forth meat and *out of the strong came forth sweetness.*'

How this riddle was solved and what happened next makes little narrative sense. It did involve Samson rowing with his new wife, and then killing thirty men of Ashkelon for sport. But the God-fearing Abram Lyle felt he could display his piety with this little quote on his tin, while more cunningly alluding to the natural product which his syrup so clearly rivalled and pastiched: honey.

I didn't mind honey as a child, on toast for breakfast, though it was always the thick cloudy clover version from Gale's, years before orange blossom or acacia were available. But for me honey was heavily trumped by Golden Syrup, and in our house there were many ways of inflaming my early desire into a life's work.

I was very proud that Betty was so accomplished at desserts, or puddings, as those who will not use a toilet insist they be called. Each meal climaxed with dessert, and Betty's marmalade pudding, spotted dick or rhubarb crumble were hugely satisfying. Supreme amongst these, however, was a quartet of dishes in which Lyle's Golden Syrup was the essential constituent.

Syrup sponge, covered with a tied cloth and steamed in a bowl, was at the pinnacle. With hot syrup running like lava over a pool of yellow custard, it was so sweet I could almost feel the enamel hissing on my teeth.

Rice pudding came next. For some this was an occasion for jam or even honey but for me it was all about how much Golden Syrup could be loaded around one dessertspoon before letting it sink into the milky rice. Then, just before all the syrup had responded to the heat and slithered into the milk, came a last minute withdrawal to suck a warmed coating of stickiness away from the top and bottom of the spoon.

Our third family excuse for syrup was served over hot dumplings. As far as I could tell these were made from the same Atora suet as the dumplings in a beef stew. I had no complaint, but why we had them for pudding remains one of the mysteries of my childhood, along with my mother's decision to tell me that we were eating pheasant when (I discovered from my brother many years later) it was actually braised and stuffed ox heart.

Neither the rice pudding nor the syrup sponge were the first choice of Twiggy, who commanded so many pages of Betty's *Daily Mail*, but at least they had some dairy element in their defence. The dumplings were just beef fat and flour, but years before we were burdened with the notion of empty calories their appeal lay in what happened when they entered the mouth, where all taste and texture blended under the eager pressure of my milk teeth.

This odd amalgamation was also found in my final application for syrup – as a coating on leftover Yorkshire pudding. For years I felt there was an element of perversion in this, and I was glad to learn later that proper folk from the Ridings also sweeten their leftover pudding, although they use jams or marmalade rather than full-strength Golden Syrup.

Admittedly, this usage may seem excessive, but I don't like attempts these days to apply language from addiction studies, physiology or religion to matters of taste which, if indulged without good judgment, become honest greed. I was not addicted to syrup. I did not ladle it into my rice pudding in search of a sugar high. I was not abusing myself by succumbing to temptation. I just loved it.

It began with the tin, and the small effort required to prise its lid open with the end of a spoon. Then came the golden green purity of the syrup, light reflecting through from the polished silver insides. Eaten neat, the sense of its suspension somewhere between fluid and solid allowed one to chew it and drink it at the same time. But I wonder if part of its intoxicating taste lay precisely in what condemns it for many food puritans today – the very fact that it was refined, indeed refined on an epic scale.

In sub-tropical lands the cane is cut by hand, left to dry, macerated, baked, boiled and formed into blocks. Then it's sent by ship in hundreds of half-refined tons, not yet fully processed for fear of being tainted by reabsorbed moisture en route.

For centuries it entered Britain through the Dickensian fog of the Thames estuary and was unloaded from clippers in Silvertown, just as it is hauled from cargo ships today. There the processing resumes through five more stages. Reheating, crystallisation, separation, cooling, and then heating again, reverting to a midpoint somewhere between solid and liquid form. Then cooled again, and canned, in that landmark design.

For those whose food arrives in an organic box covered in soil this is an unattractive process, a fine example of all that is wrong in over-refining the food we eat today. Although I also occasionally bin a rotting turnip that cost me £3.50 in order to support the organic farmer, I disagree. I think there is a brilliance in the making of Golden Syrup, a fantastic ingenuity to replicate a product which, like penicillin, was discovered by accident. The result has given so much pleasure to millions that only a hair-shirter would deny it.

Golden Syrup also taught me my first lesson in disappointment. There came a time when I was nine years old when Betty was happy to leave me alone at home for a few hours during school holidays when she went out. I never felt alone. There was a dog and a cat and Radio One, but most of all there was the larder door, the rounded brass knob of which I could now reach and turn so that with a small clunk it swung open into the kitchen. I was not an unrestrained raider, not a secret binger. I was only after a little treat.

My specialty was skimming the top biscuit from each section

of a round tinned assortment, which could land me a delicious jam ring but also a ghastly ginger nut to balance the levels. Glacé cherries for baking were pilfered too, and even the occasional jelly cube, though that smacked of desperation.

One afternoon as Betty's Mini headed towards the A21 I opened the magic door and my eyes darted around inside. The options were disappointing but when a sunbeam over my head illuminated a green and gold tin next to the tinned pineapple slices a novel idea formed. What if there was nothing to enjoy but the syrup? Was it possible to eat it on its own?

I took the tin from the larder and placed it on the kitchen counter. I assessed the swag. Something was wrong. The normally tightly sealed lid had been put back carelessly. It was clear from the brown discolouration at the rim that the tin was not as it ought to be.

Nevertheless, I took the wrong end of a teaspoon, plied the lid away, and leaned forward anticipating that golden light. What I saw was as shocking to my nine-year-old sensibility as a few frames from a film by David Lynch. My precious nectar was contaminated with the corpses of a hundred ants, lying in a mess of syrup, which had whitened, hardened and preserved their infuriating little bodies.

Gingerly, I picked the tin up and saw that there were a couple of sticky live ants wriggling underneath, and when I turned I could see their comrades' trail from the garden, beneath the gap between floor and larder door, up three shelves and into the tin. As if carrying a live grenade I ran outside and dropped it into the dustbin. I took revenge on the line of ants, wiping them out with my stubby fingers and flicking them into the garden.

There were two consequences of this nasty surprise. In the short term, I concocted a tale for Betty's benefit in which I saw the trail first and only then discovered the tin. In the long term, I behave to the present day like an environmental health inspector when it comes to secure storage of food. That which cannot be cling-filmed is screwed tight or refrigerated, and during any picnic sweet preserves which might lure a wasp or an ant remain open for mere seconds before I ram their tops back on. This has made for tense afternoons as a guest at other people's picnics, but it's for their own good. If only they had seen what I saw that afternoon.

Deep down, what most offended me about the invasion of the ants was that I appreciated the syrup for its purity, and they'd ruined it. It was because it was pure and boiled and sterilised over and over again that I preferred it to honey. Or maybe it was because I was precisely the right age to fall doubly in love with Julie Andrews, both as the sad-eyed singing nun of Salzburg and as the adorable flying nanny of old London town. Honey was not her remedy. It was a spoonful of sugar which helped the medicine go down.

Following years of research I have learned that the very first purpose of sugar was medicinal, but I knew intuitively from an early age that it was true. For to be ill when I was a boy was to go to an altogether deeper level of sugar consumption.

Neapolitan

The jackpot medical treatment in my house came with a sore throat. It was only after I left home that I discovered I have an allergy to cats, and the failure to realise this when I was young left me coughing and spluttering for my entire childhood. When it flared into a painful throat Betty had just the solution, passed down to her from the Wilson line.

Take a couple of ounces of butter, then of sugar. Place in a bowl. Mash with a fork until mixed smoothly. Give to patient. It was called butter and sugar, and I would sit up in bed like Tiny Tim shovelling it down gladly. What this has done to my arteries we are yet to find out, but by coating the oesophagus in sugary fat it probably did make swallowing easier.

For more persistent ailments I limped along to the doctor's surgery on Bromley Common. Like many doctors of the time, mine would smoke Rothmans while prodding me with his brown-stained fingers. He had a horrible probe for the ears I especially dreaded. It didn't seem to get much of a wash, festering like a wax petri dish between patients.

I couldn't wait to escape his mustard walls for the chemist's in Chatterton Road. There we would be dispensed a bottle of potion for my exclusive use by the nice-looking woman with the big nose and the large breasts which filled her white coat, my best friend's first crush.

I adored those medicines, literally spoonfuls of sugar solution to help a few grams of suspended chemical go down. It would be many years before I tasted any cherry as good as cherry medicine. English cherries were pale red and often hard with too much stone. Before Ski yoghurt made black cherry a household taste I knew nothing of its existence, and it was only by Lake Garda in my late teens that I discovered what a cherry might be if it could ripen in proper sunshine.

Other preparations tasted of banana or strawberry or peach and were just as delightful. And if I was lucky Betty would also buy a bottle of rose-hip syrup, the function of which I still don't know, but which has left me with the palate of a toothless old lady and a love of rose cream chocolates and the cheapest, common-as-muck Turkish delight.

Sweet medicine didn't stop with prescription bottles either. Tunes, Fisherman's Friends, Kendal Mint Cake, Settlers, various brands of throat pastilles – all had a putative medical purpose which could be used as a bluff by the slippery, sweet-toothed child to further his indulgence.

Despite my occasional sore throats I succumbed to my first smoke of a cigarette with a few drags of Peter Covell's No. 6 by a river in Pevensey where he was showing me how to fish. I was eight, and he was the most ungoverned boy of my child-hood, hugely likeable except when he was kicking his long-suffering mother hard on the shins or shooting a pigeon with his air rifle. How a family who lived opposite us in Bromley could be so feral I never fathomed, but if you needed a hunting knife or a double-barrelled shotgun, perhaps even a bazooka, they could supply. These untrammelled boys would have felt at home in Arkansas, yet somehow one of them grew up to be an English professor.

On that fishing trip my dominant fear was about the cigarette setting fire to my fingers, and it was a relief to return to the south

London suburbs and regain my innocence with a proper sweet-cigarette smoking session with my friend Paul Norris.

Like two old fellows outside a Greek café we spent hours smoking our way through a single packet as we dispensed them from a realistic fold-over carton. The sense of dragging on them and holding them between our fingers was very real.

It was Peter Covell's mother who got me on to bigger things, though. In a gesture a child would remember for life, she brought a copy of *The Hobbit* to my sickbed when I was very ill one week. I still have it, complete with some unsightly stains from my infirmity. This was how I was introduced to Gandalf, and not long afterwards I took my custom to a more senior shelf in the sweet shop. Sweet cigarettes no longer measured my maturity. Thereafter I would take a liquorice pipe.

The stems were shaped at the lip, with a little bow towards the bowl. Smouldering tobacco was suggested by pink hundreds and thousands, and these were the final treat when the rest of the pipe had been smoked. I remember feeling wise with one of these in my hand, blowing imaginary smoke rings. Days when I chose a sherbet fountain seemed infantile by comparison.

Eventually the doctors did to me what they did to all children repeatedly presented to their surgeries with sore throats; they decided to tear out my tonsils and adenoids. Somehow children of my age were oblivious to the dangers of scalpels slicing through their upper respiratory tract. At the forefront of our tiny minds was a much more important and dogged question. 'Do I get ice cream afterwards?'

Somehow the image had been created in a grand conspiracy between doctors and parents that a tonsillectomy was just a licence to eat nothing but ice cream for a fortnight – and there was nothing the grown-ups could do to stop us. I would like to sit with a circle of Gandalf-like philosophers now and see if it might come up with any finer example of every cloud having a silver lining. Thus I remember merrily entering the children's ward at Farnborough Hospital in north Kent utterly unaware of

15

any serious implications of an operation under general anaes-
thesia and ready to dip my spoon.

I was sharing a room off the main ward with a pale little lad
called Adam, and within one afternoon we seemed like old
friends. That evening before our operations we worked our
way through the slurry the hospital served for dinner and earned
ourselves our first scoops of an entirely new concept in ice cream:
Neapolitan. Unbelievable. An ice cream which tasted of more
than one thing. And when we woke up from our operations, we
laughed, we'd be eating nothing but this for weeks.

The next morning we were both prepared for theatre, with me
going in first. A moment later I woke up alone with a blazing
sore throat and a need to drink. I took three glasses of warm
hospital water from a plastic jug and immediately retched it all
back up with black blood into a white sink. Adam's bed was still
empty. His bedside cupboard was empty too. I slumped back into
bed and off to sleep for another hour.

When I woke my parents were there with the nurses, and
everyone was quite extraordinarily and conspicuously kind. I
looked at their faces and thought for a while. Then I asked two
questions. One: When I am able to eat can I definitely have ice
cream? There was a resounding yes.

Two: Where's Adam?

The room darkened. My father said something about it
turning out that Adam had a hole in the heart which nobody
had realised, and that the operation had been too much for him.
Nobody ever said where he had gone, or for how long, or
whether he ever could come back. It just hung in the air.

I stayed in that room for two more days, recovering, eating
Adam's Neapolitan ice cream for him. I drank to the memory of
a boy I hardly knew from a plastic beaker of the other sweet
benefit of being ill, the yellow-cellophane-wrapped bottle of
Lucozade. I haven't drunk it since.

Pillowcase

Perhaps that's what sweet stuff is in the life of the child, a substance that comes to play many roles. One day it cushions cruel blows. The next it is the madcapped celebrant-in-chief. The way its power was wielded by the twentieth-century adults of my childhood was rife with mixed messages.

I don't have a legalistic mind but one morning in our kitchen, underneath the light which looked like a boiled lime sweet, I accidentally discovered my first loophole. Betty must have thought it was time to tell her six-year-old a little more of the ways of the world. The day had come to let him know that it didn't revolve entirely around one small boy.

I remember standing in the kitchen doorway looking up at her as she picked some mint from the garden.

'It doesn't matter what you do,' she began, from out of the clear blue sky. 'But you must always tell the truth about it afterwards.'

I mounted the rusty swing next to the creosoted fence and thought about Mr Gibbs who'd put the fence up the week before. I contemplated the wonder ride he'd given me in his Commer van, with the door slid open on the passenger side, so that as we roared towards his sawdust-yellow timber yard in the woods I could have tumbled off the bench seat into the road at any moment.

As I continued to swing I finished processing what Betty had just said. In a split second I leaped into mid-air from the swing, landed on my sandalled feet, and approached her as she was hanging bright white sheets on the washing line.

'It doesn't matter what I do so long as I tell the truth about it,' I parroted.

'Yes,' she said, like Mrs Do-as-you-would-be-done-by.

At that point the memory fades, which is a pity, because it would have revealed a six-year-old boy going to a forbidden cupboard and removing a tube of Smarties he knew was there. Pausing only to check if the letter on the inside of the cap was a *P*, he would have drunk them down from the tube, chewing just enough to prevent choking, pressed out the cardboard disc at the remaining closed end, and put the tube to his eye in the manner of Horatio Nelson.

That much I don't recall, but I vividly remember the next snatch of dialogue.

Walking back out into the garden, where Betty was looking into the tree which once had a stag beetle nest, I said, 'Mummy, there's something I want to tell you the truth about.'

Betty stayed very calm.

'What would you like to tell me?'

'I've eaten a tube of Smarties.'

Her lips pursed. I went on.

'It's not wrong, is it? Because I've told you about it.'

Heavy-billing lawyers in the West End earn £500 an hour for reasoning like that. There was nothing she could say or do. Rather than saying what must have been in her mind, 'You greedy, thieving little so-and-so. That is no excuse and you know it', she said, 'You really should have asked me first because you know that's not allowed.'

Then through her gritted teeth: 'But you're a good boy for telling me the truth.'

I remember my disappointment, for in my heart I knew I deserved an almighty rollicking for an act of deviousness, and I didn't want there to be a world where smart alecks could play with language to evade a self-evident misdemeanour. I've tended to come clean about stuff ever since – which is maybe what she intended all along – and I get very tense in rooms with glib lawyers boxing with the truth, which she probably hadn't. I don't bolt my Smarties any more either.

Maybe the difficulty we have with the morality of the sweetie is that none of the Sons of Abraham, nor Buddha or Krishna were around when sweets became so irresistible. The Bible can be read to condemn pork or yeast or buggery but sugar wasn't in its orbit.

History is full of such glitches. Christopher Columbus, the man who brought the cacao plant from the Aztecs to an initially bemused Spanish court, never saw a bar of chocolate. He refused even a sip of the bitter chocolate drink the central Americans prized so much that cacao nibs were used as a form of currency. His men thought it a swill fit only for pigs.

It was three hundred years before Europeans began to mix cacoa fat back in with cacao powder, and then blended in sugar and milk to make solid chocolate. Even more confusingly the leading figures in this were all Quakers or non-conformists, stuffed shirts seeking to provide employment and foodstuffs to fight urban alcoholism. They fought hard to build ideal towns for their workers and for many years the best of them boycotted cocoa and sugar grown in plantations where slavery was still in practice. The life's work of the likes of Joseph Rowntree, John Cadbury and Joseph Fry made possible a gift so unimaginable for my ninth Christmas that I was lost in wonder.

In our house, we didn't have parcels around the tree or stockings hanging over the fireplace. We had pillowcases, flat and empty when eyes finally slammed shut after midnight and full to bursting six hours later. That Christmas dawn I went to the bottom of the bed and found an oddly shaped, tall, slim present peeking from the pillowcase which had not appeared on any wish list.

I had to wait to find out what it was until after breakfast, because my parents had pulled off an absolute coup that year with the gift of a racing bike. This had been imported from the Soviet Union, from behind what was then meant to be an impermeable Iron Curtain, by our family friend, an Uncle Derek, who regularly travelled to Moscow on errands which were never really explained. The bike had been collected from customs at Heathrow, and must have had quite a few roubles knocked off the UK price to have persuaded my father to drive to the other side of London to collect it.

We lived on the corner of a little cul-de-sac called Chantry Lane which petered into an unmade road at one end, housing a few dodgy motor-repair workshops. That morning I got used to the first pair of drop handlebars in Bromley, and to the odd concept of gears, and wobbled merrily for an hour. Only after a bowl of Weetabix was I allowed to attack the pillowcase.

The oddly upright present sticking out of the top was wrapped in red and gold paper. When it was removed I remember that I actually sat back from the gift inside and stared at it, flabbergasted that such a thing could exist anywhere but in a shop. It was a chocolate dispensing machine which, on putting a penny in a slot, would drop a tiny bar of either Cadbury's Dairy Milk or Bournville Plain Chocolate into my waiting hand.

It was the last of a kind, decimalisation in the following year making the slots obsolete. But for a chocolate-lover, and at that age I loved it fondly and without restraint, it taught the most peculiar lesson. It was on a par with men of the time building a home pub in the corner of their tiny lounges. It messed with the economy as explained on the radio by J.K. Galbraith and Maynard Keynes. It undermined the sense of what could reasonably be available at home in the way of luxury.

Above all, it taught any potential shopkeeper precisely how to eat his stock. For although I had to pay my penny to get to the purple-wrapped rectangle of Dairy Milk, I had only to open a drawer to get the penny back. It was something for nothing and, so long as a parent could be cajoled into a trip to Woolworth's for replacement bars, nothing could stop the good times rolling on.

The Band

Never in life is money spent more diligently, or with such care and awareness of consumer choice, as pocket money on sweets. In sweet-purchasing all of us had the kind of information in our heads, continually updated, which would win a job today as Chief Analyst for Deutsche Bank.

I first exercised this power at a shop opposite my primary school, next to the bus stop on Mason's Hill. The colours of my world at the time were precisely as gaudy as Austin Powers'. The sweet shop sold Yellow Submarine die-cast models and piped 'Lady Madonna' through its state-of-the-art stereo speakers. Women really were wearing tiny floral dresses, and I really did follow them up the stairs of red double-decker buses, perplexed that they were showing me their frilly knickers. If I'd been any older this might have put me off my sweets, but fortunately there was serious business to be done, for at the shop I had made a number of retail decisions.

The first was to ignore the machine outside with its horrible over-priced gobstoppers and unwrapped bubble gums. This red metal box with a hint of rust was both uncompetitive and a health hazard, raindrops or condensation running down the inside of the glass depending on the weather. It was set on the pavement between the figure of the little girl with the calliper holding her Spastics' Society collection box and the PDSA Dog with a slit in its head. This seemed inappropriate, competing with charities, and it was the sign of the world which lay ahead that over the months all three became chained and padlocked to the front of the shop.

There were two main sweet choices on the way home from primary school. In any weather other than snow an Ice Pop was always an option. Tearing the top off with my teeth, I'd suck like a new-born calf to get all the flavour and colour from the ice, then crunch dutifully through the tasteless frozen remnant to get to the mother lode. At the bottom, in the fold of the plastic, in its purest form, was the very syrup concentrate which had given the whole entity its taste. It took a while to work it up to the opening with fingers and thumbs, but as it dripped on to the tongue it was as evocative of false-flavoured fruits as my most favoured medicines.

The second choice was more intellectual, and concerned the interplay between sheets of bubble gum and the cards contained in their packet. Blighted by an early onset of facial hair, I had to manage gum with respect. One blow into a bubble too many and I'd be left with my fuzz stuck together by the product of a Malaysian rubber tree, blackening into tufts until it dried and dropped off days later.

Because of this I didn't buy the gum for itself but for the collectable material that came with it. My longest spending spree on the same brand took two months until finally I had the nine cards fitting together to make a landscape image of a painting from the end credits of *Captain Scarlet*.

The Captain, doomed yet again, but certain of resurrection because of his Mysteron-tainted blood, had been chained up, weighted down and thrown to the bottom of the sea. Looming towards him was an enormous shark who was either wondering why the Captain's smart hat with integral communicator had stayed on under water or was about to tear a strip from Spectrum's finest.

I kept these cards in a rubber band in my jacket pocket, and because I was the only collector at my school there was no market in trades, so I had to keep buying. If I win the 'Mona Lisa' at auction one day it will not give me the thrill I had on the top deck of the bus when I found the ninth card, with the shark's tail and the endless blue deep beyond. To an adult observer I must have looked like a masticating boy, gormlessly chewing his way home, but the thoughts of small boys are often much greater than they appear. Some of us were thinking about the visual arts even as pink bubbles burst across our chin.

Indeed, if I caught the red 47 Routemaster north towards the heart of London rather than south towards north Kent I passed the large, attractive art-deco factory for Robinson's Jam, set back from the main road in well-laid gardens like an idealised manufacturing unit from Trumpton. It's an enormous laundry now, but for years it produced jams and preserves, and despatched them under Royal Charter to the world beyond Catford.

Robinson's understood the interplay between art and sweetness, but its central motif could not survive the century, or even its latter decades, when the local white population moved south to the deep suburbs, as my parents had, and were replaced by a hopeful wave of immigrants from the West Indies, Africa and Asia.

It's unclear whether it was the simple fact of the motif being named the Robinson's Golliwog, or something about its woolly hair and minstrel make-up which made it a target for the Campaign for Racial Equality. Its original design intention was all innocence, but its promulgation was problematic.

To me, a child with many small black friends at school, I was not yet aware of any complications. I made no association whatsoever between what I saw as charming miniature figures and people who weren't white. I simply collected the little paper golliwogs secreted behind the jam labels and, once I had ten, I sent them off with a postal order for a shilling and received back a little statue of a golliwog in a blue suit playing a saxophone or a clarinet, part of a collectable big band.

The band stood proudly on my window sill at 103 Bromley Common and looked out as the traffic went past to destinations such as Biggin Hill and Pratts Bottom. They were optimistic, attractive and could be made to dance in a child's hands as music from *Two Way Family Favourites* drifted from the kitchen while Betty made Sunday lunch. They are about as acceptable now as a Nazi World War Two helmet. It became impossible to defend them. Yet their loss to the dustbin left a small stain of guilt on my childhood, and I wonder now if that was really necessary at all. Because without a golliwog I would not have seen my first act of true kindness.

Candy Floss

The setting scorched itself on to my young mind because it was so surreal. Cumberland is a very long way from south London and we took the train from King's Cross to get there. At that age all trains seemed like the Orient Express, their carriages alive with stewards with coffee stains on the white cotton cloths over their arms, and women in hats.

The landscape was like nothing I had ever seen, the industrial Midlands, the North. When we arrived we were collected by my aunt's husband, Alec, a hugely admired family figure who was a leading pathologist. He drove us through the Lakes in a capacious Rover which looked like a turtle, and we were reunited with the two cousins I adored, Caroline and Janet, both a decade or more older than me, and my Aunt Joan, our only real aunt.

The next afternoon we went to the fair, and there I rode my first helter-skelter, the backs of my legs burning on the coconut matting, and took my first body blows on a bumper car. Then from nowhere I was looking up at a man holding a bunch of helium balloons and noticed that my Aunt Joan was buying me one, and carefully tying its string around my index finger.

I lowered my arm so that I could see it face to face and the images entranced me. Three golliwogs in black and white suits, genial in a white oval frame, set in a brown balloon. I felt so proud of this balloon, as if I was taking a pet for a walk.

It was a clever buy because, now I peer back through the years, I realise this was not a visit to the fair for the exclusive delight of a small boy but was an adult trip to the County Agricultural Show, with a few fairground attractions added on. Most of the next hour would have been spent looking at prize marrows and polished tractor parts, but it wouldn't have mattered, because I had the balloon and my three floating friends.

The next thing I knew we were approaching the gate to go back to the Rover, parked in a muddy field, and, as we were about to pass through, the knot on my finger slipped undone and the balloon headed off into the sky. I jumped, but missed it, and Aunt Joan jumped too, but it was gone.

She must have looked down at me and seen the image of a desolate boy who has just lost his favourite thing.

'They're going for a little fly,' she said.

'Are they coming back?' I asked.

'Look, they're waving at you. Wave back. Bye bye, James. Bye bye, John. Bye bye, Jacko.'

As the balloon spiralled towards the clouds I waved back, and said goodbye to my three lost friends who my aunt had instantly named so that I'd never forget them. Then they were gone.

The face of the child left on the ground would have graced a funeral. I remember the trouble in Aunt Joan's eyes as she searched for inspiration to stop the day from ending in tears. Abandoning the prejudices of her class about the types who ran fairground stalls she took me to a man who looked like Bill Sykes and asked him to make me a candy floss.

It was a violent pink excretion of spun sugar, the centrifugal force created by a ferocious foot pump rather than the electricity of more recent times. It seemed to have pink blood spattered around it where the flavour and the sugar had not fully melded. What it produced was bigger than the balloon. Bigger than my head. One minute there was nothing on the end of the wooden stick, the next there was this eruption of cotton wool.

My face must have changed, but only to confusion at first. What were you meant to do with it?

Aunt Joan picked off a little piece with her fingers and put it gently near my lips. The first chew was woolly, the second was sweet. The third was a melting sensation, and then it had evaporated. So I moved my mouth in for the kill.

'Nice?'

'Thank you.'

I kept everyone waiting by the Rover while I licked every last strand from the stick, and was surprised to be allowed to drop it on the ground, so great was the desire to protect the burgundy leather upholstery within.

As we drove away I gazed up into the sky to where Jacko would have gone, and licked a bit of stickiness from my hand. Facing front as the car heaved through the country lanes I sniffed contentedly and looked at the back of the head of the kindest woman in the world.

Wedding Cake

My love for sweetness couldn't stay at the level of an infant for ever. It was my destiny to become the kind of boy akin to one Chrysippus of Tyana, as described by his admirer, Athenaeus of Naucratis.

Living as a Greek writer under Roman rule in the second and third century AD, Athenaeus said of Chrysippus that he 'was learned in cake'. Perhaps some day my tombstone will say that too.

For although those early recorders of the food of the ancient world were writing about honey rather than sugar, they lived when people really knew how to create a cake for a party. The most striking designs in the ancient world were reserved for festivals known as Thesmophoria, female-only occasions designed to honour their goddesses and promote fertility in both women and cereal plants.

In order to enjoy their cake the women needed to forget how they'd fertilised the grain last year – with the dug-up, rotting pig remains of a previous festival spread across the fields. If they could put that to the back of their minds they were in for a treat.

At Syracuse they invented the *mulloi* cake for their Thesmophoria, where they chose to mould sesame and honey into the shape of a woman's genital organs. From the deep id of the usually restrained Spartans came the *kribanai* cake, in the image of a pair of female breasts. Not to be outdone at the home ground of the Thesmophoria, the Athenians constructed a cake in the shape of a proud penis and testes combination.

There wasn't much call for this kind of thing in Bromley, although the Sponge Kitchens bakery could do a Formula One racing car or a Tower of London to match anyone. These were often astonishingly accurate and worked as a social triumph at a birthday party, but clearly lacked the lurid heat of the pre-Christian Mediterranean oven.

My best chance of a ritual cake was a wedding. The vast family on Betty's side kept us attending them for a decade. There wasn't much fun in them for me. Too short to see, too bored to listen, withering with rage at the centuries the service seemed to last, I saw nothing but paralysis and incarceration in my hard pew.

Things livened up for about one minute with the sublimated violence of the confetti-throwing by the church porch, but hit inertia again at the reception. The food was usually done to a pulp and the speeches oozed smutty innuendo like the latest *Carry On* film. The only reason to draw the next breath as a boy, dragooned to attend in my school uniform (because it's smart), was that the entire senseless act of tedium had to reach a climax with the cutting of the cake.

I was learned in this too, in as much as I had privileged information about its provenance. Betty was rightly highly thought of as a consultant on food matters, and if a family member asked her to recommend someone to bake a wedding cake the name of Elsie from Mason's Hill was invariably offered and commissioned.

Elsie was one of those spinster ladies so busy that it seemed I saw her every single time I left our front door. There was an elderly cleaning lady called Molly in the same league, who I later found acted as a spy for Betty when I was a teenager flirting at a bus stop with a buxom girl called Sharon. Elsie would have had too much pride to resort to that, and I felt honoured to be taken behind the scenes one day to her home kitchen HQ.

Elsie's first obstacle as a cook was to be what was then called a hunchback. When she wanted to take something from a shelf she had to swing her head round under a long arm to see what she was reaching for, but she made no complaint and, if anything, seemed more nimble because of it.

The prevailing atmosphere in her kitchen was the tang of brown sugar and rich butter carried in a mist of white air filled with flour dust. She and everything else were covered in fine particles, and when she loaded ingredients into a mixing bowl her swift actions left a hole in the atmosphere where she had just been.

So when my many second cousins rested their palms on the back of their new spouses' hands and pressed the knife down for the great ritual – 'Pray Silence for the Cutting of the Cake' – I knew where it came from and that it was going to be good. A fruit cake as rich as could be, with a thick marzipan and trowel loads of perfect icing, laid to smooth perfection and then piped to baroque glory.

My heart would beat a little faster when a waitress whisked the bottom layer away into the kitchen and a train of her colleagues came back almost instantly with newly cut slices for us all. I think there must have been something in the anticipation in my young face which always seemed to land me a corner piece loaded with masonry chunks of extra icing.

It was at this point in the day that I didn't care that I was sitting next to a dull old relative. At least we could talk about the cake. And then, as if that was not enough, we were sent away with another piece in a silver cardboard box tied with a ribbon. Perhaps some guests squirrelled this away for luck at the back of their larders, but with the turnover of fruit cake at our house Betty had no interest in curating an extra slice. When we got home it was immediately eaten with a cup of tea, and we remembered the day. There had been a nice kid called Steven and second cousin Malcolm had given me a lift in his new Capri – maybe it hadn't been such a bad wedding after all. Would Elsie be doing the cake for Maureen's wedding too?

Royal Icing

Perhaps it was her friendship with Elsie which led Betty to the Great Christmas Disaster with the Royal Icing of '72. Or maybe it was some karma for the lost life of a whale.

Whatever, Betty planned Christmas that year as always with confident ease. I can see now that she was probably tempted to ring a few changes. I'm not sure if that was the year we had a tinsel tree instead of a real one, or when we belatedly upgraded the thruppenny bit in the pudding to a 10 new pence piece. It was, though, the year that Betty declared that the Christmas cake would be covered in royal icing.

This of course sounded posh, and as I hadn't yet entered my Marxist phase I felt the same sense of anticipation as everyone else. It was much talked about, and when Des and Ena arrived on Christmas morning from the Baptist Church where Ena played the organ it was the subject of the day. Royal icing.

The first clue that all was not well came when my mother tried to mount the usual decorations on the cake the afternoon of Christmas Eve. The Santa sledge, the snowy tree and the reindeer wouldn't stand up. The problem was that the surface of the icing was like a glacier to the soft powder of normal icing.

Our traditional decorations were abandoned – a bad sign –

and, years ahead of her time, Betty had to dress the cake with minimal ribbons and a tasselled wrap like a Hawaiian skirt. The cake was placed in the larder, its astonishing fate still beyond anyone's imagining, and we set about Christmas lunch, the Queen, a short walk to the woods, and *Disney Time*. Then, in the two hours before a supper with more turkey, bubble and squeak and mince pies, Betty served a full afternoon tea.

Buttered malt loaf, a scone, and then the pièce de résistance, the royally iced cake. But as soon as it made its entrance on the tea trolley we could all tell that something was not right. Betty had removed its outer dressing in order to cut it, and what we were presented with was as plain and unadorned as the black block of nothingness in *2001 – A Space Odyssey*. It seemed to have taken occupation on the trolley and then sealed itself off from our eyes.

Basil Fawlty would not be created for another few years, but Betty's attempts to cut a slice of this cake were on a par with his efforts to withdraw a cork from a bottle of wine. Her first try was with an ordinary silver knife, more than sufficient to cut through the usual soft icing, marzipan and fruit cake. It didn't leave a mark.

Next came a bread knife, but this made an odd grinding noise against the surface of the icing and also failed to score.

'That's a bit odd, Betty?' said Ena.

Betty returned from the kitchen with as fine a blade as Sheffield had ever produced, sharpened by the knife grinder the week before. She worked it into the surface of the cake, but again it would not give.

'Give it here, Betty,' said my father, smothering his laughter and keen to get involved. Nothing. Changing tack, he realised that if he could only nick the cake somehow he would have found a weakness he could exploit. He placed the carving knife tip-down into the centre of the cake and bore his thirteen stone down heavily. The knife bent, the cake did not.

'Did you call it royal icing, Betty?' said Ena. 'Have you made it before?'

36

Ena lit a cigarette and puffed, settling in for the show.

I didn't help.

'What's wrong with it?' I demanded. 'It's like concrete.'

True, but much too blunt.

'Has it got something in it, Betty?' asked Des.

My father left the room, while Betty explained that with royal icing you didn't just use water and icing sugar, and maybe a little colour or flavour, but you beat in egg white and then, near the end, a few spoonfuls of glycerine.

Glycerine is a strange substance. In those days, it's possible it was one of the many products derived from a fin whale, harpooned by a fast boat off the coast of Norway, and turned into anything from photo-chemicals to engine oil. The really nasty thing about killing whales with harpoons is that the explosive charge is filled with nitro-glycerine, the latter part of the equation quite likely to be derived from the whale itself.

My father returned to the room. In fact, he returned to the house, for he had just been outside into the dark December afternoon and fetched a saw from the garage.

'You can't use that,' objected Betty.

'What else do you suggest?'

'Well give it here. I'll have to wash it first.'

Betty disappeared to the kitchen sink, leaving us with the sullen cake. When she came back in it was with a tool sterile enough to remove a gangrenous arm in war.

My father began to saw. There was a very small amount of icing dust but though he wielded the tool like a lumberjack he could only make a tiny groove less than a millimetre deep. This was tricky. Inadvertently he had taken the pressure off Betty and there was male pride at stake. The joke was moving towards him.

'This won't do,' he said, and returned to the garage. This time Betty had to sterilise a chisel.

'If we can just make the first crack,' he said, and placing the tip of the chisel at what he guessed would be the weak point he gave it a small tap with his best hammer. Then two medium taps, and finally three crashing blows. Nothing. Concrete would be yielding by now. What had my mother created?

My father scratched his head.

'What'll you do now, George?' asked Ena.

'Have you got an axe in the garage?' said Des.

My father nodded.

'Well I don't see what you've got to lose, do you, Betty?' said Des.

By now we were all aware that this was moving from a culinary crisis to an industrial incident, and so we moved through to the kitchen. My father set the cake dead centre on the breakfast bar and took to it with the axe. The rest of us peered through the serving hatch or around the door, wary that when the cake did crack there could be shards flying through the air.

We needn't have worried. The axe would have blunted before the icing would give way.

'Why don't you drop it, George?' said Des. 'It needs some kind of all-over impact, doesn't it?'

Both men worked in motor insurance and theoretically had a lot of knowledge of how things crashed and buckled. My father agreed with Des's idea.

The problem was that it needed to be dropped from the bathroom window upstairs, but if that worked we'd be left with dirty wet cake all over the crazy paving. So my father raised it above his head instead and threw it with some force onto the kitchen floor. Just as Des predicted, the overall stress of this caused fractured veins to open all over the icing cover, like Lake Ontario thawing on the first day of spring. At last the surface was cracked.

Unfortunately, so were the floor tiles under the kitchen lino, and for years we could tell which ones had taken the brunt because of a slight crunching sound when we walked across them. But we did at least have access to the innards of the cake.

More than one guest that Christmas wore dentures and they were not about to risk them on the icing, so they just ate the still delicious cake with whatever marzipan could be salvaged. I tried the icing, only to discover that it was harder than the most over-cooked pork crackling. I sucked a piece for a while, but Betty said I should stop because if I accidentally swallowed some it would choke me on the way down. We never had royal icing again.

Angelica

It's a phenomenon of this half-born age of mass media that we don't always encounter art of great fame in the form in which it was conceived. In my early childhood any outings for classical music on the stereogram were of the James Last plays Beethoven kind. All the hooks and catchy bits were segued and the dull Germanic context left on the recording room floor.

Similarly, I first knew *The Thousand and One Nights* tales from a cartoon featuring an Americanised Ali Baba. But I was not the first cake-lover to take this Oriental hero out of context. When the Polish king Stanislas Leszcsynski was thrown into exile in Lorraine, he found the traditional *kouglof* cake too dry, so he reached for the rum and splashed it on all over. He too liked bowdlerised Ali Baba stories, and so the rum baba was conceived, and then propagated by French chefs of the nineteenth century competing to make the very best.

I first encountered this rum syrup soaked marvel, with its cherries, angelica and cream (which I scraped away and left) at a changing point in British café history. The traditional Lyons-style tea shop was struggling both to pay high street rents and to compete with the rise of burger joints like Wimpy's or versions of the US coffee shop such as the Golden Egg.

Betty would not, as she often said, be seen dead in those places, and until I had some independence in my teens I had to rely on reports of maple syrup waffles, brown derbies and frankfurters from friends of more voguish homes. I spent years glimpsing ketchup dispensers shaped like tomatoes through steamy windows before I had my first squeeze.

The most common time for Betty to take me out for tea was after school, when we would take the bus back into the heart of Bromley. There, the afternoon reached a fork in the road. Depending on her mood, or who she thought we might run into, we would go to one of two restaurants. The Coffee Importers reeked of freshly ground coffee. I've never been able to swallow a single mouthful of the stuff, and the acrid caffeine air was too hardcore for a child. I could have overlooked the whiff if they had something I wanted to eat, but their non-savoury range was very limited, and I'd usually write it off as a bad job and ask for Welsh rarebit with a glass of lemonade. At least I could have some fun making noises with the straw while Betty nattered with Audrey or Dorothy at the next table.

The alternative was a much more welcome choice. Betty might have a small amount of food shopping to be done at David Greig's in Market Square, perhaps breaded ham on the bone (to be had with a soft-boiled egg, peas and Branston pickle later). On leaving the shop with its polished black steps and gilt lettering, straw-boatered staff fighting the final battle for independent grocers, we faced the main door across the High Street of Bromley's equivalent to Arding & Hobbs or Bentall's, Medhurst's.

It would be wrong to say that Medhurst's had faded grandeur because it had never really been that grand. It had character, though, and from its rear windows on the top floor you could look down on to what little of the Ravensbourne Valley had not been built on, towards Shortlands and the faintest hint of central London in the very far distance.

Before they installed escalators we used to go to the top floor in a lift with two doors. One was a brass grille which would have stopped a rhino, the other a solid sheet of metal which would have killed him. Both were liable to slam small fingers in their hair-triggered mechanism, so when we reached the roof-level restaurant it was always with a beating heart. In those days Betty had a fur coat, and though we were hardly Michael Winner and his millionaire mother entering the Carlton in Cannes, we did sweep in with a certain élan.

These were years of my life to be treasured. Before long the restaurant would become self-service, and Medhurst's rebranded as Allder's. But in those last years of the great department store teatime, I would watch my rum baba carried across the room as if it were a whole lobster in the professional hands of a waitress wearing a black dress, white pinny and lacy cap.

The glass dish was round with low sides, so spoon work had to be surgical to avoid sending the baba skidding on to the floor. The first incision was like carving the best piece of roast beef from a rested joint, the outer layer where flavour is concentrated at the edge and the texture is all sealed softness within. As ever, I'd dump the cream, chew blissfully on the glacé cherry and tickle the roof of my mouth with the angelica on the tip of my tongue.

Like the hard-to-handle glycerine that caused the Hunka Munka debacle with the royal icing, the angelica also had a surprising origin. It too was raised in Norway, like the fin whale, a form of bamboo distantly related to sugar cane, which when arduously and multiply processed – including one long three-day reboiling session – could be candied and used on cakes. In Norway children take school bread with them in the mornings adorned with the angelica.

Intuitively, if Betty and I had been asked to guess its place of origin (in the days before anyone asked such questions) we would have said Morocco or somewhere deeply Ottoman. Its proliferation around the world following the Viking plunderers serves to show that heavily processed food has been with us longer than we might have guessed, and that we might sit back for a moment before thinking it is all a conspiracy of the modern age.

Wagonwheel

I like to think that, despite some later disagreements, Betty and I were happy in each other's company on those afternoons, as were my father and I when we went out as a pair. I'm going to have to pin down a name for my dad because although his work friends and wartime colleagues called him by his second name, George, he was christened Basil, a name only Aunt Joan used.

Betty, however, long before my time, had decided to call him Peter. When I first read of Christ in Galilee renaming Simon who shall be called Peter, it made more sense to me than anyone else in class. Peter, then.

Like my mother's, Peter's life was shaped by the war. As a tall, skinny artillery man he had crossed the Rhine at the age of twenty-five and found himself staring into the faces of Belsen. He hardly discussed this with a soul but it had given him certain views about race and Europe – liberal values in a conservative carapace. I once asked him the typically child-like question of had he shot anyone, but he ducked it. He probably hadn't but the cannon he targeted at Nemegen probably had.

Much of the horror which he was exposed to as a young man was grimly predicted by the most famous man ever to have lived in Bromley, H.G. Wells, though like Scheherazade's *A Thousand*

and One Nights I saw him first through the wrong end of the telescope. I knew him as the screenwriter for lots of Vincent Price films. I did not then know that Wells was born and raised in Medhurst's when it was a mere draper's, nor that the wild-looking man Peter and I saw once a year at a football match was his distinguished biographer, the Labour minister and later leader, Michael Foot.

The site of our annual encounters was the rickety wooden west stand at Charlton Athletic Football Club. In the war years, when Peter first became a supporter, Charlton was one of the mightiest clubs in Britain, but if the perception is that England was succumbing to a sickness in the seventies, then the virus had a head start at the Valley, an old gravel pit with a capacity of roughly 85,000 people which struggled to get 5,000. Consequently, the stand was so empty that you could sit where you wanted, and Michael Foot used to bring his family to support his beloved Plymouth Argyle whenever they came to Charlton.

Probably hoping to get noticed, Peter led us to the row behind them. Come the revolution he might have been the first to go, but if the rebels had come from the Orpington Conservative Party then Peter would have directed his own rifle at Mr Foot and Anthony Wedgwood Benn, who he sincerely believed to be the credulous toff, dupes of communism, reds under the bed.

With unconscious irony, given his estimation of Mr Foot's rhetoric, Peter would feed me humbugs throughout these often dire fixtures, and I'd leave a shaming pile of wrappers under my seat at the end of a game. We didn't say much, and I must have been frustrating company because I had no sophistication at all with regard to the run of play. Goals were all. I never noticed the work of the midfield or the defence. I only had eyes for my idol, the striker Derek Hales, who many years later would play a key part in both my football and cake-consuming life.

From the point of view of either the dentist or the nutritionist (or the Pavlovian behavioural conditioner) Peter was doing me few favours with this one-every-ten-minutes humbug tactic. It was only when I took my four children to watch football that I remembered what an eternity a game can seem to a restless child and, despite the once-a-week sweetie regime in operation in our house, I resorted to a sack of sherbet lemons when I took them to endure Exeter City playing Charlton in a recent Devon friendly.

At least I would never give them what Peter fed me on a particularly freezing winter's night at the Valley. The visitors were Crystal Palace, then cocks-of-the-walk, managed by Malcolm Allison in his fedora and sheepskin. As ever, we could sit wherever we wanted, so placed ourselves just in front of Allison who was sitting in the front row of the directors' box.

I was quivering with the cold and with excitement at being near someone so famous, taker of team baths with Fiona Richmond, father of white bling. Peter, like Betty, had his culinary standards, but, seeing how cold his son was, he let them slip and at half-time he went to fetch me something to hold body and soul together as the wind whipped through us.

He returned with those disastrously mismatched staples of the football match, a cup of hot Bovril and a Wagonwheel. It may be that Heston Blumenthal could do some molecular wonder with this combination at the Fat Duck. On my sweetened palate they fought like cat and dog.

The Wagonwheel has somehow sustained a place in British confectionery right up to the present day, even though nobody admits to eating them. The advertising campaign, centred on trapped cowboys under wagons with arrows through their hats fighting off injuns and being rewarded with one of these immense circular objects, was always persuasive. Their continued success is a mystery. The chocolate is so thin as to be like a brown paint, the biscuit seems stale, and the marshmallow inside looks like a squirt of shaving foam. That it sold then and sells now can perhaps only be attributed to price sensitivity. They may be crap but for your money they are an awful lot of crap.

For hot Bovril there is much more of a defence. Rendered ox carcass, anyone? Like gelatine, it went vegetarian after BSE, but on a cold day it was meaty, salty, sweet and aromatic. It could go with bread, cheese, Ryvita biscuits, a tomato sandwich, indeed just about any savoury food in Christendom. Why football caterers decided to sell it with Wagonwheels will never be properly explained.

I was a few bites into my Wagonwheel in the second half when the referee blew the whistle to call one of the Charlton players offside. Peter said under his breath, 'That was never offside,' so I leaped up and shouted in my unbroken voice: 'That's rubbish, ref, he was well onside.'

To my amazement a smooth cockney voice came into my left ear.

'Sit down, kid. You don't know what you're talking about.'

It was Malcolm Allison. I was livid.

'Yes, I do,' I squeaked. 'He was through on goal.'

'But he was offside.'

'No, he wasn't.'

'Do us a favour, kid. Keep it shut.' Then he gave me a small clip around the ear that Peter didn't see.

There were tears in my eyes now. I hated him. I hated him because he was talking too harshly for an adult addressing a child. And I hated him for trying to say that my father had got it wrong.

Peter stayed quiet, but I wasn't done. Like an offended injun I had my eye on Allison now. A few minutes later he turned round to talk to a Crystal Palace director. I seized the moment to flick the remainder of my Wagonwheel up into the air over my head towards him. It landed on the floor between his feet. I then sat with my heart beating for the rest of the game, waiting for him to drag me by the scruff of the neck and give me a good duffing up.

He didn't, and at the end of the game when I turned round to look at him he smiled indulgently and ruffled my hair, unaware of any outrage he had caused. As he walked away from his seat

48

and down towards the dressing rooms, though, I was both thrilled and terrified to see that stuck to the bottom of his crocodile shoe was my half-eaten Wagonwheel.

Yee-haw!

Pockets

There must be millions of us of a certain age who waved our primary-school friends goodbye by Christian name one summer afternoon and arrived at secondary school two months later to find that everyone barked at you by your surname. My childhood caught a chill the day I discovered I would now be known as Arnott. Even bosom chums called me Arnott for the first five years of our acquaintance. It presaged a harsher reality in my mollycoddled life.

The actual voyage to school changed too. A two-minute bus ride to a smelly but kind little primary school became a three-quarter-hour journey by bus and train towards inner London, followed by a walk around the roaring South Circular Road.

A seamless transition to my new school was accidentally hindered by Betty's decision to take my first-day haircut into her novice hands. She bought some scissors, a hair thinner which scraped out your follicles, and a salon-style sit-under hairdryer. For technique she was guided by an instruction pamphlet which came with the thinner. I'm not at all sure that this wasn't designed for pets, and am certain it was used on our cocker spaniels later on. They shrank away and growled too.

Betty's first cut was the deepest. Finding it hard to balance her handiwork at the back of my head she took it up an inch here, then an inch there, until finally I was sent to meet fifteen hundred new schoolmates with a haircut I didn't see again until the first series of *Blackadder*. At primary school I had been pretty popular. In my first weeks at Dulwich College, with its insane catchment of hard-nut free-place kids from Tulse Hill and softboys from Petts Wood, I was easy pickings.

I'm hoping this might evoke some sympathy, because my coping strategy for all this was not what was expected of a fellow from a school whose alumni included Wodehouse, Shackleton and someone we could never remember who'd won the George Cross. Like an urchin, I survived by finding my consolation in small familiar pleasures. My junior-school sweet dealer on Mason's Hill was now in the past. I moved quickly to establish a new supply chain, and was thrilled to discover that I would be able to fix myself at both ends of the railway journey.

At Bromley South station there was an old-fashioned kiosk, where one lady stood up all day in an area of eight square feet growing varicose veins. Later on she would sell me my first Rothmans.

At West Dulwich station, six stops north, there was the Nook, an isolated hut with a corrugated-iron roof run by a bearded man and a woman who looked like Golda Meir. His fingerless mittens failed to conceal his yellowing fingers which he'd dip deep into a jar of sherbet lemons, dropping them stickily into a four-ounce paper bag. Somehow the first few sucks always tasted slightly of his Old Holborn and the cat who sat on his wife's feet, a hint of a lemon-scented kipper.

The shock of Big School was offset, to an extent, by my taking command over my life as a sugar consumer. Like a junior Micawber, however, my books didn't really balance, and this was exacerbated by the unforeseen financial implications of Betty sending me in with a packed lunch, rather than my eating what was served in the dining hall. On Mondays when she made me a roast meat sandwich from leftovers of the day before I was fine. Taken with a Penguin biscuit and a satsuma, I was a contented and appreciative child.

By Wednesdays Betty resorted to Bovril and cucumber in Mother's Pride slices. Unfortunately, the way the white bread absorbed the cucumber's moisture during the morning made it fit only as rolled bread pellets for the birds by lunchtime, and usually ended up in the classroom bin. I'd put my lunchbox on a radiator until break-time, and if it was egg and salad cream (always with a tiny fragment of embedded shell) I needed to ensure I disposed of it outside or be accused of letting off a stink bomb in the afternoon.

Effectively going without lunch three days a week, I was genuinely hungry. The solution was sweets. The problem was money, especially as Smarties were now sold in boxes. One tube didn't seem worth opening my mouth for when a box secreted in a jacket pocket allowed greedy fingers to reach in and palm orange ones into my mouth all day. My goal was to get hold of at least two boxes a week, but the reality was this would not be covered by my pocket money. Too cautious to steal them from shops, I decided to steal the money to pay for them instead.

Motive, Opportunity, Method, that's what all the detectives said then. Motive was hunger and greed. Opportunities, without which there would have been no Method, were two-fold – Betty's handbag, and the hanging jacket of a boy whose father worked at a Middle Eastern embassy in London.

Method a): Check Betty's whereabouts. Go to side of bed where her handbag is stashed. Examine purse. If near empty, take nothing. If bulging with coins take one or two. Never take a note. Be careful about 50p pieces. Assume that only Scrooge would know how many 10p pieces he had, and remove swiftly.

Method b): Play rugby. Everyone into horrible muddy baths. Don't take long in bath and slither like an eel back to pegs. Identify Middle Eastern boy's jacket. Feel weight of pockets. Check for observers, dip hand, remove loot. He always had so much money it seems highly unlikely he ever noticed. I really hope he didn't because, bar stealing from him to feed my sugar lust for a couple of years, I rather liked him.

There is no excusing the dishonesty of this, though I can see how much I immured myself to the strangeness of school by giving myself a hard sugar coating. Even the journey was fraught. As the years went by we learned to slash the seats and the string luggage racks of the rolling stock with our Swiss Army pen-knives. We'd unscrew the lightbulbs and throw them on to the track, pee copiously from the window, and draw genitalia on the walls in felt-tip pen. And that was the swots.

My best friend fell off the back of a bus one day into Bromley High Street and could have ended up as strawberry jam. I fell on to the railway tracks at Sydenham Hill as a train shot past on the parallel line. We were regularly robbed at knifepoint by a boy trading under the name of Eric from a very rough neighbouring school. There was an odd teacher called Robo who carried nail scissors and snipped off what he referred to as our sideboards if our hair grew over our ears. What he did with these raped locks is a matter he took to his grave.

Then there was Latin. And Physics. Teachers like hippies, and teachers like Nazis. Over-sexualised boys up to all sorts who, one reasons now, must have picked this up at home, inflamed by the goonish, nudge-nudge, wink-wink voyeurism of the period's low-rent film and TV. And at the bottom of the heap, poor, sad, good-as-gold children called Simon, who I'd travel back in time to by Tardis and scoop away from the dumb hurt of it.

There was nothing I could do about school life until I was older, so I made my own world of pop music, sport and confectionery. As the years went by I added a third supplier too, Mr Sansom in Locks Bottom, who was the last of the sweet-shop-owners for whom selling confectionery was a vocation.

Old Sammy, as he was known, had such a white complexion that we wondered if he was an albino. A more probable reason for his geisha look was that he never left his shop in daylight, appearing from a darkened room at the back like a guinea pig emerging from a hole. He spoke precisely like the comedian Max Wall in a nasal staccato, always kind and courteous, but desperately shy. You might be a child but you were still a customer, and owing to the extreme gloom in the shop usually the only one.

He always wore a black suit, white shirt and dark tie and could have won a job as funeral director anywhere in the world. He can't have been eating his stock because his tiny collar was always loose at the neck. One might imagine they broke the mould when they made Old Sammy, but in the shop next door his brother ran a clock-repair workshop and jewellers. Seen together they seemed like prototypes for Gilbert and George.

Unusual characters like the Sansom brothers are riveting to children. I would go into the shop marvelling that I lived in an age when new sweets seemed to be launched in a stream of constant innovation and try to crank up a conversation about the debut of the Old Jamaica bar. But though his shop reminded me of a dark quayside chandlery where a pirate such as the one on the wrapper would come for a new cutlass, Old Sammy himself always held his own counsel. He neither approved nor disapproved of his sweets, they just were. His dry-as-dust persona might have discouraged a child, but his willingness to sell you a mere ounce of sherbet strawberries if that's all you could afford won our respect.

Many of the great new names I would have loved to discuss with Sammy have disappeared. Mint Cracknel, Opal Fruits, Spangles, Aztecs, Texan bars (The Mighty Chew), Pink Panther bars, Lord Toffingham ice creams, the range of Cresta sweet drinks (It's frothy, man).

Some survived. Curly Wurlys, which once removed teeth, can now be bitten through by new-born babies. Love Hearts still act as an instrument of courtship on the top deck of buses. Freddo the Frog is half the frog he used to be, but still appeals to children who like to eat body parts in consecutive order.

My mouth was like an industrial laboratory for new products. With the gravity of a don discovering the double helix I found that you could combine one cube of Cadbury's Dairy Milk with a single mint Tic Tac and then suck and suck and suck until they were all gone and you still had more cubes and mints left in your pocket. If they'd let me study the chemistry of that at school I might have been a scientist.

Beet

I still remember the first lesson when we were taught of the infernal triangle of eighteenth-century Atlantic trade: slaves from Africa to the Americas, sugar from the Americas to Britain, guns and trinkets and goods from Britain to Africa. Then repeat.

There were two black boys in my class then, recent immigrants from East Africa. It might seem unlikely now, but I hope they'd remember that, at least within school, they never encountered any form of abuse based on the colour of their skin. Their integration was complete, and any difference in identity respected. The horror of the historical fact of sugar slavery didn't seem to affect them, or at least they didn't want to talk about it. Then, the ABC mini-series *Roots* was shown on the BBC and presented what happened to Kunta Kinte and Chicken George in the name of slavery as prime entertainment on a series of Sunday evenings.

It was 1977 and everyone knew the races had to be nicer to each other. When Reggae Sunsplash or the Rock Against Racism concerts took place in nearby Brockwell Park we were all enthusiastic (though to be honest I preferred the Electric Light Orchestra). But our institutions were so hopeless. Kingsdale, a largely Afro-Caribbean school next to ours brought their steamingly fine jazz band to play a concert in our great hall one lunchtime. Their rip-it-up rendition of the *Hawaii Five-O* theme tune was the most exciting live music we had ever heard. Two weeks later we returned to Kingsdale with Dulwich's own musical offering – a string quartet playing Bach to a music teacher and six boys in detention. We were deeply ashamed.

Playing a friendly game of cricket that summer against Rastafarians on Clapham Common, our opponents refused to borrow our batting pads, so that even the slowest dolly from our superb Anglo-Australian bowler kept splintering their shins. We kept offering, imploring even, but they still said no. A few years later I was on a train at Brixton station looking down into Acre Lane as the riots seethed below. All of it so obviously of a historical piece with the brutes who shackled young Africans into ships two hundred years earlier, and inflamed by the perceived injustices of the day.

Yet at that time we were encouraged to believe that the consequences of the early sugar trade were all over, that with the abolition of slavery in the nineteenth century we had done the right thing. I would go for long walks or cycles to a large area of green-belt countryside where we lived near Keston and Downe. Downe was the home of Charles Darwin, who in the years before the publication of Richard Dawkins' *The Selfish Gene* seemed to have a much lower public profile, to be merely historical. We associated him with stuffed animals and the smell of the biology lab at school. His house lay unvisited, unmarked, practically uncurated.

But that same cycle ride also took me to the Wilberforce Oak, near Holwood House, which had been the home of the Prime Minister William Pitt in the late eighteenth century. William Wilberforce recorded in his diary of 1787, later inscribed on a stone seat by the oak tree: 'At length I well remember a conversation with Mr Pitt, in the open air at the root of an old tree at Holwood just above the steep descent into the vale of Keston, I resolved to give notice on a first occasion in the House of Commons of my intention to bring forward the abolition of the slave trade.'

I could lean my bicycle against the railings protecting his tree, just a blown core then with a newer tree emerging from its old trunk, and imagine that the thoughts originating from this same spot excused us the wickedness I had seen in *Roots*. Yet even then I never believed it was the whole truth. It was just too simple. I later discovered a great work, *Sweetness and Power* by an American academic, Sidney W. Mintz, and had it confirmed that the black and white issue of the abolition of sugar slavery remained grey for decades. The commodities market made that a certainty.

After abolition the British by and large did not want to handle or consume sugar produced by slavery, though places such as Manchester were not as committed to seemly good practice as London. But forty million Germans were even less scrupulous. They weren't concerned where their sugar came from, and the British merchant navy realised that, as long as it wasn't for sale in its own country, it could still carry cargoes of slave-grown sugar to Prussia from Brazil for great profit. Indeed, they sometimes refined this sugar en route in London, so that the Germans and the Brazilians didn't have to bother, and then invoiced both countries for this noble service.

Then they could either process the payments through British banks, or ship back to Brazil from Germany a payment in kind. If the British refined too much sugar for the Germans to consume and were left with a surplus there was no problem. Of course with our Christian consciences we could not eat it in Kensington, but we could ship it on to the colonies, to Australia or even the West Indies themselves, where it could be consumed by criminals or natives or both, who were assumed not to be so troubled by moral niceties.

William Wilberforce, therefore, was the beginning of the end of slavery, but given the current policies in world trade, where the First World countries subsidise their own beet and dump any excesses on the international market, many Third World farmers' lives are little improved today. My own theory is that the man who made the abolition of sugar slavery possible was not even English, and it certainly wasn't his intention. He was before his time and he didn't know it. He made possible a new era of sugar plenty for our own times, while rendering the use of slavery to produce it redundant.

Franz Karl Achard (1753–1821) was a German chemist who carried out experiments to see if sugar could be derived in large quantities from a product other than cane. Beet was his answer. Under the patronage of the French and then of the Prussian King Frederick William III, he invented the industrial process for the extraction of beet sugar.

This is best told in the infallible handbook for all small-scale sweet manufacturers: *Skuse's Complete Confectioner*. In amongst its diagrams of lozenge-cutting machines and recipes for cough candy lies the following passage.

The beet industry at first languished and would probably have expired had not Britain's naval blockade of Europe in the Napoleonic Wars cut off the supply of cane sugar to the European continent. The high prices then realised for beet-sugar enabled the struggling industry to survive and expand. This increased production materially lowered the prices of the commodity and incidentally brought cheap and wholesome confectionery within the reach of the masses. At the present day sugar is no longer regarded as a luxury but is recognised to be an important staple foodstuff, and a cheap and valuable source of carbohydrates in the most quickly assimilable form.

It was because of Achard that the sugar refiners began to source raw sugar on their own doorstep, cane sugar becoming more of a luxurious item. Many food puritans of today may hate it but this beet sugar kept millions alive across Europe during the terrible privations following the industrial revolution. It feeds my beloved Golden Syrup factory today.

Achard, in the grand tradition of men who change the world, died broke before he could see the fruits of his work. And Holwood House, where Wilberforce made his declaration before Pitt, was throughout my childhood a training centre for employees of Tate & Lyle.

I used to sneak into its grounds as a child, and it was once a stopping-off point during a whole-day snog, aged fourteen, with a girl called Deborah. Heaven knows what tips she'd gleaned from *Jackie* but she kissed me from Tugmutton Common to the High Elms Golf Course via Keston Ponds on an extraordinary five-mile walk. Halfway through we bought Strawberry Mivvis from a hut by the ponds amongst the tree roots and the frogspawn, a fine blend of artificial strawberries and vanilla chilling our hot breath.

Vice

My father made up a nickname for me which he used until he died. He used to call me Vice. This came from my appointment as vice-captain of Form JC, aged nine. Even when I attained the full captaincy he would still call me Vice. It left me with an eccentric interest in the nearly men of America – Hubert Humphrey, Walter Mondale, Spiro Agnew, Dan Quayle, Al Gore et al, though with the ascension of Dick Cheney its appeal evaporated.

Peter had another description for me which I liked less. He'd take a swig of tea, look at me fondly and say, 'You're just a cupboard love, Vice.'

I didn't like this for two reasons. The first was that I was his son but I was adopted as well and, though I loved him, I felt uncomfortable if I felt that was being tested. I was already carrying a small burden and preferred to get on with it without having to prove my filial devotion. The more telling reason I disliked it was that he was right. After all, I was a boy who endured weddings on the promise of a slab of icing eight hours into the day.

This love of my subject, sweetness, pulled in two directions at once. It could be a strength, giving negotiating power over the course of a long day-trip from home. But it was a weakness, making me biddable, prone to manipulation. If I most resembled any animal it was probably a dog. I wouldn't have chased sticks, but I did have an empathy for a hound who'd jump through hoops of fire for a Doggy Choc, the prospect of which made even the smartest pooch into a dumb mutt.

Peter had the most peculiar commute to work by now. His insurance company had moved to Folkestone on the Kent coast, entailing a three-hour round trip every day. I don't imagine he enjoyed this much. For him, the main consolation of being stuck in the commuter cycle was the proximity of our home to the south-east England countryside. But to a boy with his ear glued to the wireless, learning the world through Radios One to Four, the countryside was even duller than a wedding.

Years before gay life took it on and gave it a sheen, 'popping down to the Lanes' in Brighton was a cripplingly awful prospect. As was a day out to Hever Castle, Penshurst Place, Knowle Park, Sissinghurst, Battle Abbey, or Chartwell, and I shuddered at the mention of an oast house. These days, an old fart, I'd enjoy a day out to any of those places, but then I had to be dragged from home on one very strict fully negotiated condition: what are we going to eat?

If we were going to Eastbourne for the day the following T&Cs would apply. At the earliest opportunity we would visit Bondolfi's, a grandiose seaside tea room with something of both the chocolatier and the pâtisserie about it, without really being either. Set back from the front on a large, open, florally-embellished roundabout, it always had a fine display of chocolate rabbits, footballs, and an enormously proud white-chocolate cockerel.

I would take my place facing these visual delights like a pilgrim looking up at Our Lady of Lourdes and ask for a Rose's lime cordial and a rare beef sandwich served 'club-style' with crisps and pierced with a cocktail stick. This too was part of the negotiation – you couldn't have afters if you didn't eat your befores.

While my parents had something creamy, I took a one-sixteenth chunk of a chocolate gateau, powdered with cocoa and filled with aromatic and well-chilled ganache. Bondolfi's was the first time I leaned into cocoa and breathed it in, the smell of sweet earth released through the refinement of an extraordinary pod.

This small rite of indulgence bought my parents a couple of hours' peace to do whatever adults did on the shores of the Styx. Maybe listen to a brass band pumping out 'Without You', meant to be a secret song about my teenage heartache not a hum-along for oldies under tartan rugs.

With my flares and safari jacket billowing in the breeze I'd head off to Eastbourne pier, drawn like Pinocchio to the coloured light bulbs. Fuelled and thrilled still by the cake, I took to the one-armed bandits, and shoved pennies into slots, hoping to tip a cascade of coins over a teetering copper cliff.

Funds depleted, I bought a stick of rock and topped up my tan on a bench facing west. I held the thin white label in my palm and studied the black and white photo of the pier over and over again. The pink rock consumed my complete oral attention, until finally I'd confirmed that this one read *Eastbourne* all the way through too, and I rested my aching jaw. Rock was the first foodstuff in my life which I knew had to be finished where it belonged. Like ouzo or a holiday romance, rock did not travel well. Then I'd look at my watch and meet my parents by the entrance to the pier. It was time for an ice cream.

I was glad to see them again, asking them what they had been doing but never listening to their answer. I was focusing on making sure that I was the purser for the acquisition of the ice cream. Left in charge my father was likely to get a disastrous rectangular Wall's slab wrapped in paper to be placed between two rectangular wafers. This was a catastrophe, and though it might have reminded him of his own childhood he needed to remember who was the child here. My mother could not be trusted either. She'd been known to return with the same slab stuck into a cone with a squared-off mouth.

Only I understood the national mood. It was to be a 99. Nothing else would do. As the years went on this became a double 99, and then a double 99 dripping with squirts of strawberry-flavoured sauce.

As with the paddle of the Kenwood Chef, I displayed an innate sense of drama. First, I had to devour one of the Flakes. Then lick away the melting ice cream from the cone's edge. With a stiffened tongue I forced the remaining Flake down into the stem of the cone. Then I bit off the end and sucked ice cream through the hole with all the force of a baby at the teat.

Returning to the cone's edge I nibbled away the parts now sodden with melting vanilla ice. I licked it out and, moving at speed, sucked the teat end again. Thereafter all was abandon, a loss of control as I devoured the remains. I sighed, licked my salty fingers, wiped my hands on the back of my trousers and asked my parents when we were leaving.

A Tray of Pastries
(and a Toasted Teacake)

As one grows older there is something appealing about the customer-is-always-right character of an American diner. Pull over to the side of the road in your hired Pontiac, get wished the warmest of good mornings, and dream up a bespoke order from the menu. If you want sunny-side-up egg in the centre of a frosted doughnut covered in BBQ sauce you only have to ask.

The proper English tea room, a disappearing institution still to be found on cathedral greens and seaside lanes, was never so relaxed. The place we used to visit on our way back from Eastbourne was stiff with starch, and scared local girls serving ladies in hats out with their retired brigadiers for a run in the MG.

It was a twenty-minute spin inland from the coast on our route home and we'd pull over into a car park set to the side of the road in the Ashdown Forest. Like so many things from childhood the name of this wide-open, heather and gorse landscape was completely confusing, in the sense that there wasn't a tree to be seen.

As I grew older I was excited to learn that the forest had been denuded centuries earlier to make timber for the Royal Navy, but

I didn't care about that then and leaped out of the Triumph 2000 with Tina the cocker spaniel and sprinted off down the tracks through the rough cover, imagining that I was on an army mission. If I was organised I might even have brought my Action Man and his armoured car in the boot with me for a bit of rough and tumble in the sandy soil.

Then we'd run back to the car. Before crossing the deserted road to what I genuinely think was called Ye Olde Tea Shoppe Betty would ask me to tidy myself up. On entering, the wildness of the treeless forest fell away, and one was overcome by a sense of propriety. Everything was in its place. Nobody wore a baseball cap or a football shirt. We were dressed for tea.

As I became a teenager I rather fancied myself in an atmosphere like that. Fresh from the sea, tan topped, young muscles beginning to ripple, I used to eye the young waitresses with an unwholesome vanity. Recalling that now it seems it was probably a reaction to the sheer sexlessness of the place. This was no Bondolfi's, which was a bordello in comparison. I had Hawkwind's 'Silver Machine' drumming through my head, while everyone else in this Tudorbethan experience had more of a Glenn Miller vibe.

What I could never take away from Ye Olde Tea Shoppe was that it made the best toasted teacake in the south-east. Their teacakes were unusually huge, almost too big for the plate. As if some Italian mamma had let rip with the dough, they weren't symmetrically round, and they were lavishly buttered. Just a few decades ago, heavy buttering was considered generous and good for you, and margarines were only fit for baking and getting big toes out of bathroom taps.

These teacakes were always too hot to hold on arrival at the table, and the butter slithered on to the plate down its sides and through the heart of the beast itself, leaving a circle of yellow every time I raised it to my lips. I learned to love drinking tea here too, slaking my thirst with a steaming Darjeeling and plunging the teacake into my mouth before I had swallowed it.

I was grateful to the toasted teacake too, because it made me one with my family in this shadowed, half-timbered, mullioned room. It allowed me to be different whilst also being with them, for when the waitress approached, either immune or slightly set back by my lascivious gaze, my mother always had the same order.

'A pot of tea and a tray of pastries, please.'

To which I was permitted to add and, once my voice had broken, in the tones of a Capital Radio disc jockey, 'And a toasted teacake, please.'

For the tray of pastries was my enemy. All the items on its doilies were stuffed with cream, which I loathed. Wherever we went, but particularly in a summer tea room, the cream seemed flyblown, a clotting agent, repulsive. My loathing meant that I shunned the glories of the chocolate éclair, the cream slice with its fabulous wet icing, and the cream sponge.

Sometimes the day was saved by confectioner's custard, when the cream slice was transformed into the vanilla slice. If I saw this on display in a Sussex window I urgently requested it as an addition to the tray of pastries, but this was all too rare.

Outside Tina was tied up lapping a bowl of water, while inside my parents enjoyed what they sincerely considered one of the great traditions of English life. Of course, Betty would be all smiles with the waitress, and then tear into the cleanliness of the knives, the smeared windows and the wobble of the table legs the second she was out of earshot, but this was part of the ritual. To her and my father, motoring from London's fringes to the countryside and taking English tea was a luxury.

Perhaps all children are unappreciative of the things their parents value, but as I grew older and heard stories of pilots during the war motoring down with their girls from Biggin Hill and going down blazing the next morning it grew special to me too. My godmother was one of these women, eventually dying fifty years later, still wearing her young man's wings, as faithful to him and as unmarried as a nun.

Was there something in this tea which really was unique to the south of England? Certainly, elsewhere there were local specialities associated with parts of the British Isles which did not feature at all on our region's bill of fare. The Eccles cake, the Bakewell tart, lardy cakes, barmbrack, saffron cakes were all indigenous to the islands, but on menus in Kent and Sussex they featured as rarely as zabaglione or tiramisu, while the toasted teacake was the bedrock on a tea-room menu, its supreme incarnation arriving in Easter week when it rose again as the hot cross bun.

After tea we returned to the Triumph and before we had gone too far out of the forest my mother offered me the final sacrament of the day. She opened the glove compartment and unscrewed the lid from a tin of car sweets, a dry spray of icing sugar rising into the air. Car sweets were to be sucked not chewed, and it was a matter of self-respect to get back from mid-Sussex to Bromley on no more than two. In any other context these aged sweets which lasted years in the car would have been roundly rejected, but their crude cherry taste seemed perfect as our car fell into line along with hundreds of others returning from the coast to face the working week.

As we drove on I balanced the latest copy of a book from the Jennings series by Anthony Buckeridge on the dog's back as she dozed across my lap. These schoolboy comedies were as perfectly crafted as a Wodehouse. The fictional Linbury Court was set in the same Sussex where I'd spent the day and, though I knew nothing of boarding school life, I could imagine the thrill of being taken out for tea by a kindly aunt on something called a half-holiday, of looking forward to Exeat weekends. Above all, I understood the moral outrage at another chap helping himself to your tuck box.

Buckeridge was far too fine a writer to resort to japes with prank sweets, as found in the *Beano*. One of his best stories perfectly captures the comedy of consequence in confectionery as he describes Jennings and his chum Darbyshire spending a genial afternoon scoffing a box of chocolates sent out of the blue by Jennings' aunt. Only when it is empty do they open the accompanying card which explains that this expensive assortment is meant as a gift for Matron to thank her for nursing the school through an outbreak of chickenpox. The chums are horrified, but with no postal orders due for a month they have to rely on their own resources to fill the empty tray with whatever they can find. Luckily, Darbyshire is a hoarder and still has the original wrappers. Over an agonising week they wrap these around lumps of sugar, cough drops, bubble gum, indigestion tablets, barley sugar, doggy chocs and anything else that tastes remotely sweet. The scene in which Matron shares the reconstituted box with the teaching staff as J & D look on, especially when the terrifying Mr Wilkins gets the Rennie stomach settler, always made me howl with laughter.

Years after I first read the books I was ferried by coach from south London to have my head kicked in playing rugby at all sorts of boarding schools in south-east England. At Christ's Hospital in Horsham we were bewildered to share our pre-match lunch with an entire dining hall dressed in cassocks. I tried to talk to them about Jennings and his assortment box, but by then they had turned their heads against childish things and were much more interested in the latest double album by Genesis.

The Falls

The first time I studied my entire body was one early morning in the Swiss town of Interlaken. We didn't have full-length mirrors at home, certainly not ones I could linger in front of stark naked, and so with the sound of cascading water roaring through my open window I was able to look at myself in a Swiss wardrobe door. I was fourteen years old and making my first visit 'to the continent', a holiday that began a lifelong and unfashionable love of the Germanic countries at the heart of Europe.

At that age, no longer a child but with eyes still wide open, my imagination was captured by every new sight: Alpine houses, cheese factories, oompah bands, the fountain in Lake Geneva, my first glacier, the colour of girls' tans in the high mountain sun, gay men with moustaches walking hand in hand in Zurich station, roads with no barriers and thousand-foot plunges through the clouds.

All this left the impression of a country in which the people were not cowed, as we were in strike-bound, IRA-bombed London. Switzerland was where Burton and Taylor had flown to avoid taxes, taking with them her azure eyes and the shapely bottom I had seen briefly in *Cleopatra*. Switzerland, I considered, was a place only the jealous said was boring – to me it seemed to break out in music and sport and laughter like cuckoo clock-work.

I was old enough to go off on my own as early as I liked, and when I woke in the sleeping Hotel Mittel I quickly dressed, centred my parting and went to explore Interlaken. The only street action came from vendors opening their kiosks filled with risqué magazines, cigars, and chocolate. I had enough phrase-book German to make myself understood.

'*Bitte, geben Sie mir ein Lindt*' were my first words to a foreigner in his own tongue. Holding the bar lightly so as not to make it melt, I walked away down the empty street, wet from the street cleaners, and found an immaculately kept park. I sat down on a bench and undressed the first foreign chocolate of my life.

With the after-dawn sunshine heating my pimply forehead, I lay the Lindt on my lap and slipped it from the foil. It tasted odd, but fine, already softened in the morning heat, tasting slightly of cheese, but if flattened to the roof of the mouth and compressed with the tongue still full enough of cocoa and sugar to please. I dropped its purple wrapper in the empty bin and returned to the dozing hotel, all its wooden shutters still closed to the new day.

My room had a single full-length window which opened inwards and let in the ceaseless roar of a waterfall in a gulch outside. The view was dark and filled with trees, and I had slept only fitfully the night before, forced to keep the window shut to dampen the noise. For the first time ever I had slept without pyjamas, the temperature in the sealed room staying in the high seventies all night. Now that I'd returned I stripped off again, intending to nap in the hour remaining before breakfast.

I wanted to sleep, yet my biology teacher had told us that if we slept immediately after eating we would store all the food energy as fat. I looked at the naked grey reflection of myself in the mirror. Never having studied myself in my birthday suit in

Bromley it was like attending a life-class where the specimen happened to be me.

There was no fat. I had grown more than a foot in the preceding four years and any fat had fuelled the growth of bone and muscle. I could see the undulation of the lower ribs in a disproportionately large ribcage. My hips were narrow, and were balanced on the sprinter's thighs. The effect seemed longer and slimmer than I realised, and holiday photos show it was the only time in my life when I might be described as skinny.

As the torrent beyond the window shook the room I looked at my flat stomach with its tiny trail of hair and I realised that this body of mine would change one day whatever I did, and that it would be better to obey my instinct to sleep rather than to stay awake growing bags under my eyes for fear of putting on half an ounce of surplus fat from the Lindt bar.

I understood that the butterfly season cannot last for ever. Germanic Switzerland had taught me that too, a race of blond, blue-eyed people who by the time they reached the age of soaking nakedly in the hotel sauna had grown guts substantial enough to hide their genitals, their hair grey and their fading eyes sharp behind wire-rimmed spectacles.

That waterfall mirror spared me from any future agonies of food anxiety, anorexia, bulimia, self-contempt, self-harm, from straining to body-build by steroids or killing myself in the gym. It was the moment I accepted the physicality of the Fall, and it came in a place just like a Carl Gustav Jung consulting room. Once understood, it released me to enjoy my food, to accept the body's changes as they came. But it also lowered my guard at seeing the dividing line between simple indulgence and idle force of habit.

Years later I stood face to face in mirrors with partners as they assessed the two reflections to see if we fitted. I watched their eyes wander over aspects of themselves they would change, that they disliked, but despite my sincere assurance that their cellulite was perfection their discontent was real and brooding. Isn't there a kinder word for 'sag'? Isn't the way we succumb to gravity part of the Fall, and isn't the grace with which we fall the best and only ride on the planet? How do you make a Swiss roll? You push him down the mountain.

Strudel

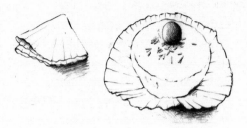

That Swiss trip was responsible for showing me there was cuisine beyond Bromley, even though Switzerland didn't have a reputation for fancy cooking. I was like a country lad with all London before him but dining only at the Swiss Centre. To the novice palate of a teenager everything Swiss was exciting, even sauerkraut, or cold plattered slices of cheese and ham – for breakfast! Accompanied by dark heavy breads with names like pumpernickel, full of seeds and grit.

Was this the German bread that Julie sang about in *The Sound of Music* in the 'Doe-a-deer, a female deer' song? 'Tea,' she warbled, 'a drink with German bread.' It was years before I realised she was singing 'jam and bread' in that deceased diction you can't find now in even the prissiest voice class at RADA.

Even walking the Swiss streets literally took my breath away as I fuelled myself at the kiosks on small strange packets of mints. These released icy fumes in my mouth on entry, and after a couple of bites numbed the gums like a dentist's gel preparing you for the big needle. I couldn't find these mints in London for years, until I stumbled across them in the diabetic section of Holland & Barrett during my brief dried-fruit phase. Sugar-free sweets did not compute.

It was one Swiss cake discovery, however, which changed my perception of what could be done when baking with the simple apple. In America I'd be condemned for dissing my own mother's apple pie, but the truth is that in a wide range of delicious foods it was the only dish Betty never really mastered.

Her pastry was long and dry, heating through but doing nothing in terms of texture, merely encasing the apple. The fruit itself was roughly sliced and not marinated or trickled with lemon or honey or cinnamon or anything else to make it taste good. It was as if she had hold of a recipe from a thirteenth-century monastery, so plain and unadorned was the outcome. It was made actually dangerous by the profusion of cloves she threw in, which ended up impaled in your gums or down your gagging throat, serving only to make the apple pie taste of disinfectant.

On a coach outing through the Bernese Oberland one after-noon, we stopped at a village and ate outside a Bierkeller. By night it hosted Swiss dairy workers singing 'Trink-Trink' and by day it fed trestle-tables of white-legged British tourists. We were served an inclusive meal which my parents did not much want. They weren't really coach trip people, and the company made them feel even more à la carte. However, the meal was fine, beginning with a potato soup, followed by a juicy German sausage with chips, a vinegar-doused salad and a mild mustard which could still make the eyes water.

As we looked across a busy road to the range of mountains we had driven through, a procession of wonderful-looking Swiss women, who looked to me as the Elves must have looked to Sam Gamagee, filed from the kitchen into the burning sun. With shimmering tongs they served us each a precise slice of the cake which made it possible for me to eat cooked apple again. This was my first encounter with strudel.

First impressions last. I knew this one would be with me for life when I saw the icing sugar sprinkled liberally all over it. It sweetened my initial disappointment at realising this was some kind of apple dessert. It was made sweeter when I held the plate up and saw the filling was plump with raisins too.

The pastry was like nothing I had known in Britain. It was thin, like moist filo, made with a well-worked, hard semolina flour, doing something with pastry that we simply didn't do. Instead of lining a pan with the stuff, loading it up with filling, and slapping a lid on top, a baker carefully winds his strudel around whatever is within. (Strudel means whirlpool or vortex.) It can be cheese or plum, but mainly it is apple.

I could have tasted something more accomplished in Hungary or Czechoslovakia, but they were still behind the Iron Curtain. Vienna would have been the best place (at Demel's café – a *meisterwerk* served with the world's finest custard) where it had first been brought by invading Ottomans hundreds of years before. However there was still a chill between our nations after the matter of World War Two, so I found my first strudel on a Swiss coach trip.

The saying is the same wherever in Europe it is made: 'The uglier the apple the better the strudel.' This idea was counter-intuitive in late twentieth-century Britain, but in this millennium we've begun to understand that we should not judge an apple by its skin.

When I returned to school that September I tried to explain to my friends how fantastic this apple strudel had been, but in the context of other people's holiday romances, real and imagined, and the fact that the Parkers were the first of us to have been to the exotic new destination of Florida, it all seemed irrelevant. It was twenty years too soon for me to find anyone of my generation with whom to hold an appreciative conversation about cake. In south London, baking still meant the manufacture of millions of white sliced loaves at the Tip-Top factory near Orpington, where everyone who ever worked there emerged with the same story of the night shift pissing in the dough.

Marlon

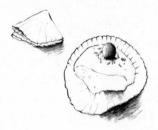

The subject of the body image promoted by the media to women has been under discussion for many years. Less discussed is the body image sold to young men. My teenage generation was given the most perverse role models to emulate. It is unsurprising if many men don't like the way they look, having been raised in this time of confusion.

First there was the matter of bodily hair: bad or good? Sean Connery had the lush chest of an ape and Bond girls happily twisted their fingers in his doormat. Roger Moore, on the other hand, in *Live and Let Die*, removed his shirt and revealed the puffy pink nipples of a harem eunuch on a sunburned hairless torso. Yet Roger's Bond girls stroked his smooth chest too, as if their lives depended on it. Couldn't they see that he hadn't reached puberty?

Destiny carried me down the hairy route. In school changing rooms jealous smooth types would occasionally quip after Esau in the Bible: 'For Arnott is a hairy man.' But I always knew they'd have killed for just a few follicles of what I had.

Yet all around us now men seem to be plucking backs, sacks and cracks as if being hirsute is a source of shame, and then sharing with us in interviews that 'Victoria likes my balls hairless'. My theory is that it's only men unable to muster more than a few coils of nipple hair and a sparse off-centre tuft between the ribs who have grasped at this straw. To me it is as unnatural a development in civilised society as implanted J-Lo buttocks.

Then there were the concepts we were sold regarding our musculatures. As a child all my war comics had small ads with sand being kicked in the face of weedy men, who returned to the beach the next year with their chests expanded by Charles Atlas's patent flexing apparatus. Then in a rather unchristian way these new Titans duffed over their former persecutors to the dancing joy of women in bikinis. Some of my school friends actually bought these hopeless machines.

But this was as nothing to the impact on our young psyches of Arnold Schwarzenegger bursting into the seventies, rippling like a waterbed in his teeny pants. Like all his body-building fraternity Schwarzenegger had taken the triangular shape my mother had adored in Kirk Douglas and Robert Mitchum and hung some seriously pumped meat on it. Although Schwarzenegger's inflatable army always had skinny calves, and looked as if one gust of wind would send them weebling off down the road.

My mother's generation had not expected to see actors with their shirts off, but if you catch an old clip of Clark Gable or Spencer Tracy in the jungle you'll see their trousers are pulled high in the Simon Cowell style, little paunches held in by strong fabric and a belt. They were as puny as everyone else. Arnie, and then Sly and Jean-Claude set the bar to maximum bicep and made even the chest-flexers feel puny.

Fortunately, most of my school friends were too cool to buy into any of this nonsense. Cruelly, we laughed at the boy who pumped so much iron that when he fell over backwards, forgetting to let go of the weights, he broke both wrists and came to school with his two arms plastered in front of him like a cartoon. Or the poor cadet, desperate to join the RAF, who spent every lunchtime building a body like Hercules and was turned down for failing the eye test.

Punk rock exploded around us when we were fifteen years old yet, despite Siouxsie and the Banshees, the Damned and many more being from our own home turf, we turned our noses up at it. Bowie had already come from our patch, and they were crap compared to him. We already knew punk was a marketing trick. So, cleverness blights the freedoms of youth.

To us, most great contemporary art came from the America none of us had ever visited. Very early Springsteen. Lou Reed. Lynrd Skynrd. Dylan, of course. Early Eagles. The Doors. Poets of the airwaves. As for physical role models, we rejected cheesy old farts like Charlton Heston, and looked to what we thought of as the unassailable oeuvre of Marlon Brando, though admittedly our actual knowledge of his work was patchy.

This was based around two phenomena. We were doing *A Streetcar Named Desire* for O Level, and in a retro shop in Soho you could buy black and white stills of Marlon in his ripped shirt yelling 'Stella' up the New Orleans staircase after the disappearing wife who wanted the beast in him in her. This was much more exciting to us than Larry or Johnny or Ralph. This was an actor who people wanted to have sex with.

What we didn't know was that this image was already two decades old, and that in the real world Marlon was getting ready to wear the worst wig in film history and play Superman's father. To us, Marlon was about big biceps, ripped red t-shirts and leather jackets. That was the look for me.

So I felt a keen teenage sense of destiny when it was announced that my school was going to mount a huge production of *Guys and Dolls*. I was seventeen years old and hadn't done much acting, although with proleptic irony my first line on stage had been as a Sheriff in Henry IV describing Falstaff as 'A gross fat man, as fat as butter'. Like every other eager auditionee I heard that this was a production with girls. Auditions were being held in various sister schools across south London. There was to be mingling of the sexes.

With beginner's luck and to the chagrin of those boys who'd

sweated their way up the Drama Society through Shakespeare productions, declaiming poetry and sucking up to the cravat-wearing, self-styled Director of Drama, I was cast in the role of Sky Masterson.

Though none of us had ever seen the Warner Brother musical of *Guys and Dolls*, word spread round the school that I was playing the Brando part. The kudos of this could not be measured on any normal scale, fifteen hundred boys further inflamed by the rumour that I would have to kiss the deputy head girl of James Allen's Girls' School who had scooped the part of Sister Sarah.

That Marlon had been simply dreadful in the film had not occurred to any of us, even twenty years after its release. He'd sung as if his chunky jaw was wired and was devoid of that prairie innocence Damon Runyon intended for his character, the sweet poker face which made Sky such a great gambler and a hit with the dolls.

Yet my connection with Sky went much deeper than Marlon, beyond the coincidence that I looked exactly like the man Runyon described as having baby blue eyes and a round kisser (once I'd sacrificed my sideburns for my art). Because Sky's very first lines in the libretto are about Nathan Detroit's attempt to bet Sky that he did not know whether Mindy's Café on Broadway sold more cheesecake or more strudel.

The director of this production was called Reggie Cowling, a fair-bearded Lawrentian English teacher of some genius with a beautiful wife and a slightly camp love for Broadway musicals. Although he turned out to be a bit of a Bob Fosse when it came to staging, the English teacher in him made sure we grasped Runyon's language and milieu first. We had to understand, for example, why it was funny that the gigantic gangster Big Jule told Lieutenant Brannigan of the NYPD that he came from East Cicero, Illinois, and that he was a scoutmaster.

So I was proud in one of these linguistic sessions in Reggie's domain (in the form room of English 6th, with its old armchairs and sketches of vaginas behind locker doors decorated like a hippy's camper van) to shake off the suspicion that I'd only been cast as Sky because I was tallish by giving a potted account of the provenance of the strudel, a cake still unknown in our part of the Borough of Southwark.

As I spoke I began to bluff correctly that it was a staple of the New York diet imported from places like Switzerland by hungry immigrants. I could smell the cinnamon apple fumes at the back of my nose as I talked and felt as if I had made a real connection with America from a school room in sight of the Post Office Tower.

That production of *Guys and Dolls* left its mark on me in many ways. On one side of the balance sheet, the scale of my involvement in my final year at school cost a couple of grades at A Level. On the other side I did get to kiss Sister Sarah on and off stage, and that changed us both. And somewhere on both ends of the scale, it was the first time anyone ever said something to me about putting on weight.

His name was Dr Peter Buckroyd, an inspired English teacher with a theological bent and the musical director for the show. In the middle of a dress rehearsal one afternoon, performed to about eight hundred surly schoolboys, Dr Buckroyd suddenly screamed out loud like a teenage girl while he was conducting and I was singing.

The orchestra carried on in shock, and so did I, already in some difficulty facing these sniggering louts and crooning 'I've Never Been In Love Before' to Sister Sarah's understudy. Meanwhile Dr Buckroyd kept on unaccountably squealing and smashing the music stand with his baton.

Afterwards I sought him out, and demanded that he explain why on earth he'd started screaming as I was singing

> But this is wine that's all too strange and strong
> I'm full of foolish song
> and out my song must pour

which was already hard enough for a seventeen-year-old before his peers.

85

His pale face reddened and his small fists clenched.

'Because, Arnott, you skipped two bloody verses.'

From then on I hugely liked Dr Buckroyd, and I can see the fear in his eyes now as he tried to get a full orchestra to catch up with this conceited youth on stage. After the show was over we used to have the odd chat around the school. One of them casually drew blood.

'I notice you've been wearing that big grey jumper this term. You want to watch out. The bigger jumper is the first refuge of the fatter man.'

I was astonished. He had to be mad. Certainly he wore thick glasses.

'Are you still doing as much sport?' he asked.

I was cross at that. He should have known all too well that I was like some kind of unpaid PR for the school. When I wasn't warbling in musicals I was showing prospective parents around the school as head boy or attending meetings of the board of governors. Or sweating the French Revolution in the library. Prematurely, the ceaseless motion of my childhood and youth had completely stopped. I was nearing eighteen, and had started to walk everywhere like an adult, and worry about things like an adult too.

'Yeah, loads,' I replied.

'Well, just watch yourself, that's all. Remember what happened to Marlon Brando.'

As I walked away from him for a quiet cigarette behind the scout hut I completely failed to understand the reference. What was he on about? As far as I knew, what happened to Marlon was that he became the world's most famous actor and the envy of every man alive. I had no idea that to Dr Buckroyd's generation he was the complete joke from *Last Tango in Paris*, buttering some poor actress's backside while he called her a filthy pig, keeping his top on so you couldn't see what a pig he'd become himself. And I certainly didn't get that to Dr Buckroyd this self-regarding role was so obviously a sneak preview of

Marlon's inevitable physical decline. He'd seen something in me like Marlon – and with the vanity of youth I took it as a compliment. Nobody would have predicted the scale of Marlon's tumble, the buckets of ice cream. But Marlon ignored warnings too.

It was his tragedy that when he played Stanley on Broadway in the early fifties he would have been hard pressed to find a decent ice cream anywhere in New York. The turn of the century had seen America obsessed with the novelty of the ice-cream parlour, but after the war a new generation of consumers had new habits formed by the universally available freezer compartments of their mammoth fridges. Now they could take slabs home with them and share slices across the family dinner table, served with mom's apple pie.

Unfortunately for Marlon Walt Disney was about to cause an explosion in ice-cream consumption which is still reverberating today. In 1955 he created the first Disney theme park at Anaheim in California featuring Main Street USA, a cod-historical representation of a simpler American time. Looking as ever to access the collective unconscious of his customers, he remembered the ice creams of his childhood and opened a colourful replica parlour on the street. Unexpectedly it was a commercial sensation for the Mouse, and this success was noted by manufacturers across the country. Soon, ice-cream parlours were reopened in every town, and a terrible trap had been laid for poor Brando. The industry realised that the simple home slab was just the beginning, so the fake Scandinavian Haagen Dazs entered the market in 1960, followed by the fake hippy Ben and Jerry in 1978. Before long they provided their wares in family-size pails, which nestled nicely into Marlon's lap of a lonely evening ready for him to dip his big paw into. The horror, the horror.

Later that *Guys and Dolls* year my own Sister Sarah had a party at her house. Her friend Katy took a photograph of us both. There must have been something wrong with the camera because it seemed to suggest in my laughing profile looking down at my girlfriend that I had an incipient double chin.

A week later Sister Sarah said what a lovely photograph it was and could she keep it. And that I'd look good with a bit of extra weight because I had quite a big build.

Harrods

One of the most repulsive experiences of our childhoods had been supping third-pint bottles of warm sour milk forced upon us every first playtime at primary school. Mrs Thatcher the Milk Snatcher, in her earlier guise within the Department of Education, had removed this from a school's obligations and, though it probably left a generation with calcium-deficient bones, we cheered her for it. No longer would Mrs Tierney stand over us in her midi dress smelling of mothballs until we'd drained every last drop. While I wouldn't have dreamed of voting for her, Mrs Thatcher seemed briefly a good thing. How was I to know then that her government would loom over my life from the first day of my gap year to my third and fourth children's second birthday?

On the day I left school in 1979 I had no idea at all what I was going to do. The Head of English had taken me to one side in the spring and told me to my amazement that the Headmaster was going to wangle me some kind of place at his old Oxford college – a quid pro quo for the hundreds of school tours I'd led instead of using my study periods to learn some grade-enhancing facts.

When at the Leavers' Assembly, however, the Headmaster thanked me for doing the final reading and then enquired in complete ignorance as to my future plans, I realised that the crucial call had not gone in. He had not shared port at High Table and bigged me up to the Dean of Brasenose College. Indeed, I'm not sure he ever knew he was meant to, one of the perils of attending a school with two hundred in the sixth form and management systems from before the war. Too polite to mention that I'd been led to think my fate was in his hands, I ducked the Headmaster's question and stopped being a school-boy.

It was a very difficult time, for relations at home were becoming strained too, and I was about to be hit by double trouble. My parents were vehemently opposed to me becoming the first person in their family to go to university and refused any financial help. Unfortunately, my father's declared income to Bromley Council meant that I wasn't going to be able to obtain more than a few hundred pounds from there either. If I wanted to go into higher education I would have to work most of my year off and save as much as I could.

However, these earnings were not taxed at source, as happens with gap-year students today, so when I left the Personnel department of Harrods and worked out on a bus ticket what I would be taking home each week I was amazed. On the day I collected my first pay packet, an envelope filled with notes rather than a direct payment into a bank account, I had that universal, you've-never-had-it-so-good moment which has given hope to young workers for decades.

For the naïve from south London working at Harrods provided a haphazard finishing school, like an Advent calendar with windows into the lives of those who had not been raised in the suburbs. I was known as a 'Mobile', one of a crack squad of employees entrusted to work in different departments each week or even each day and do what had to be done.

Within weeks my social perspective broadened enormously. I had attracted the interest of a princess from Yugoslavia, a number of gay men from the Way-In clothes department, and the Head of Personnel. I had met young men and women my own age with every accent in the country, and in the oddly troglodyte atmosphere of the tunnels beneath the store and the vast Trevor Square warehouse every disability and race too.

In those pre-Aids days I agreed to take over a cash register in the Luggage department one day after its chatty operator whispered in my ear that there was something he urgently needed to do. Touching my hand lightly he told me he needed a minute to go and change his tampon, a remark I was still processing on his return from the lavatory freshly plugged.

The Window Dressing department, with whom I spent an extraordinary week, declared themselves to be 'the most influential poofs in Europe' and may well have been right. Having made sure my girlfriend met me by the staff entrance one evening, they became warm friends, forever ribbing me that I was 'on the turn' and didn't know what I was missing.

Harrods was a thousand different places to me, from the day I covered for the ex-school friend who'd just vomited his Christmas lunch into the delicatessen cold room to the Machiavellian back-stabbing of the Harrodian Dramatic Society. But with a strange sense of destiny I was drawn again towards the sweet stuff.

My first experience was helping launch the brand-new Jelly-Belly counter selling a new line of imported jelly beans from America in the Food Hall. We were allowed to taste the entire range so that we could comment on them for customers, but only before the store opened its doors. For breakfast I would have mouthfuls of jelly beans tasting of watermelon, pina colada, coconut and strawberry cheesecake. At the end of the day the underside of my shoes had developed a second sole half an inch thick made of squashed jelly bean, which I scraped off with my fingers in one long strip and put in the bin. Had it not been for the fluff and dust mixed up in it I would have given it a go.

Then I was drafted into a peculiar front-of-house role at the Georgian Restaurant Tea Room on the top floor. This is what the restaurant at Medhurst's in Bromley would have been like if it had a million-pound refurbishment. The tea room dripped chandeliers, the grand piano tinkling to the sound of 'My Way' played by a Royal School of Music student in one of the circles of Hell.

My job was – and of course I had never heard of it – to be a kind of maître d' at teatime, slowing down the flow of arriving customers and snapping my metaphorical fingers at a highly experienced head waiter to take these people to their tables. I couldn't believe that I was getting away with it, and felt surely that the waiters must resent me, but no, I was the man prepared to stand in a lounge suit being smooth with Americans and Arabs and they respected my labour as much as I respected theirs.

Harrods was open for six days a week and shut at five o'clock on all but two of them. Consequently, the Georgian Tea Room was empty by about a quarter to five, and I was beckoned to join the head waiter and the pâtisserie chef at the best table. With the piano player not clocking off until five he was able to jazz it up a bit.

All the waiters looked Italian, but were from all points between Ecuador and Poland and sincerely proud of their work. Their pride was validated by this massive perk, that they could sit down with Harrods monogrammed plates and a silver pastry fork and help themselves to any number of items from the long table covered in cakes which the customers had not bought.

I never tired of this ritual, and when after a month the full-time maître d' returned from illness I left the Georgian Restaurant with a heavy heart and a heavier footprint on the scales. A daily feast of marzipan- and chocolate-covered delights followed by my old staple, the rum baba, was not being burned off by a slow walk from Knightsbridge to Victoria station. It was during that month that I first felt my trousers tightening and my 28-inch trouser band starting to pinch.

I went for a long run out towards Biggin Hill one evening and was puzzled by pains down the front of my shins, by a breathlessness I had not known even after the hardest rugby match. I had joined the workforce but lost my condition, and didn't know how to get it back. I needed to devise a diet plan.

What I came up with was so idiotic that it could probably have been a hit if I'd published it. On television, KP peanuts were being advertised as 'having more protein than a plate of roast beef'. Similarly, the Milk Marketing Board were promoting the benefits of full-fat milk. Eureka. I had a plan.

The working day at Harrods was punctuated by three break times. So, at the short morning and afternoon breaks I ate one small packet of KP peanuts and washed it down with half a pint of fresh, chilled milk. This, I calculated, would allow me to push the boat out for lunch.

At lunchtime I would leave the store to sit in Hyde Park, stopping at an Italian sandwich bar for roast beef and cucumber with salt and pepper on white, followed by a custard tart the size of a side plate. And a 'low-fat' strawberry milkshake. I would then sit on a bench overlooking the Serpentine feeling very mature reading the pre-Murdoch *Times*, or feasting on *Midnight's Children* by pre-fatwa Salman Rushdie. As I nibbled delicately at this modest meal I relished every word of the scenes evoking sweet and pickle factories in India, not knowing that I was beginning to fatten myself up as surely as the greediest uncle from Bombay.

Oscar

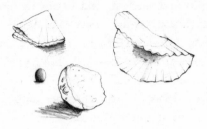

For some months at Harrods I had no idea what to do next. I wondered if I should enrol on their management training course and become a departmental buyer. Certainly some quite cool people did this, but once too often I heard talk of junior buyers hooked on free trips and staff discounts and not getting out of Harrods until it was too late. And for every buyer of linens from Milan there was another sourcing light bulbs and radiator valves. I also lost confidence in the Harrods brand when, helping my parents buy a cocker spaniel from the pet department with my 25 per cent staff discount, the poor thing proved so inbred as to be completely mad and had to be put down.

A better use for my staff discount was the book shop. I'd read the Bible, and I'd read Shakespeare, so I thought I'd work my way through the complete works of Sigmund Freud. These were conceived, although I could not then have imagined it, in such fabulous places as the Café Landtmann in Vienna, the perfect and enlightening synthesis of cake, hot drink and free thought. Today people sit with laptops and show each other blueprints on PowerPoint. In Freud's time they would have sat with cigars and talked and talked, so much so that he eventually twigged that he could make a name for himself by medicalising this talk and calling it psychoanalysis.

For the first time I found myself with an independently discovered intellectual interest, and began to look into how I might study Psychology at university. My A Level results were enough for places at either Durham or Exeter, and on the basis that it would be warmer in Devon I chose Exeter. (It snowed during my first summer there.)

My near future decided, I was then persuaded that if I went straight to university without seeing at least some of the world I would be a dull student. With an old school friend working at Harrods and two girls also in their year off we conceived the most tightly budgeted European adventure ever undertaken. I'm ashamed that it was probably my own desperation to spend as little as possible so that I could pay my way through Exeter which gave it such a miserly tone.

The advantage of this penny-pinching for me was that the weight I had gained at Harrods tumbled away. As we slowly worked our way by train through France, Italy and Greece we lived off bread, wine, fruit and, as a luxury, salami. I was the most awful hectoring figure in those beautiful markets as we pooled our eighty pence each for daily food. One really sweet and funny girl seemed desperate to buy things I had never heard of, such as brie cheese or avocado. Repeatedly I vetoed her. I now realise that she was probably entering the early stages of malnutrition, her body calling out for some good honest fats.

The only luxury I agreed to was a pot of wonderful Bonne Maman Confiture de Fraises which made a bright beginning to every morning, smeared on French bread or on a croissant dipped into a bowl of *chocolat chaud*. It was as distantly related to the Robinson's Golliwog jams of my childhood as the Picasso museum by the Mediterranean at Juan-les-Pins was to the Dulwich Picture Gallery. I sought it out for years, and now that it's widely available in Britain I adore the Confiture de Figues, a superb fig jam delicious on anything from brioche to a hot, buttered crumpet.

As we meandered down the Soane and the Rhône and into Italy I was meant to be taking in the world's great art – a full immersion coming when I threw myself rather than a coin into that fine example of the baroque: Rome's Trevi Fountain. At first I hadn't realised I'd injured myself but when we boarded the ferry from Brindisi to Patras I took off my trainers to expose what looked like the Elephant Man's right foot.

It was unclear what had happened, and I spent the next few days hopping about with my good friend carrying my rucksack for me. Eventually I was confined to a cheap hotel in Athens as my chums visited the Parthenon and the sights of the ancient world. These stifling days were made longer by the hotel also being a lively brothel, the air full of the grim climaxes of short-legged men with moustaches.

My friend's grandfather was the Mayor of Kalymnos in the Dodecanese, and he thought it best that we make it to the island and seek medical help when we arrived. All that remained was a ferry from Piraeus to an island almost kissing the coast of Turkey to discover nearly a week after the incident that I had two broken metatarsal bones in the foot. However, I was plastered, fixed with an odd metal clamp under my heel on which to hobble, and then, as my companions ate lotuses, confined to barracks with my friend's grandmother trying to make headway into someone's copy of *The Idiot*.

After a few days I could bear my confinement no more, and went hobbling off through the cicadas to find the nearest shop. *The Magus* had been in vogue for at least a decade by then, so I had the expectation of finding something weird. What I never imagined was that after a quarter of a mile of hopping and limping I'd stumble on a small, dark shop which sold the most enormous Toblerones I had ever seen. At Harrods they sold impressive lengths of the stuff, but somehow on this Greek island they had bars big enough to club a man to death with.

I had always been in awe of the triangular hunks forged into one sensational serration. Sealed in yellow cardboard packs, decorated with snow-capped peaks, they seemed to be mined from caverns under Mont Blanc. Like so many great chocolates the Toblerone never needed to look for its identity beyond the name of its confident creator, Theodor Tobler. He absorbed the word 'torrone' (meaning nougat) into his surname, and began to sell his bars through vending machines on Swiss railway stations. He conquered the world with his unique chocolate triangular prisms without firing a single bullet.

I never did find out why these mega-bars were so plentiful on hot Kalymnos, or why they were so cheap, but I had no choice but to buy one and carry it as swiftly as I could back to my hostess's house. My friends might not be back again for hours, so I decided to try one triangle to see what it was like. It was fabulous.

Granted the freedom of the kitchen, I went to the fridge and carried two bottles of beer and the Toblerone to the terrace. Within an hour the whole lot had gone and, as the sun began to fall towards the horizon, I felt as burningly ashamed as if I'd spent the afternoon rolling about in rude magazines.

I disposed of the evidence, and when my friends arrived back with talk of hunting sponges and eating sea urchin in clear, warm bays I held my peace. Naturally they were hungry, so my friend's grandmother presented a huge meal of fried calamari, stuffed pepper, and moussaka. I was obliged by good manners to eat all this, and when I went to bed that night I felt like a goose who'd force-fed himself into pure pâté.

If that day stood alone in that sojourn on Kalymnos then so be it. However, the following days all followed the same rhythm. Friends leave for warm waters, I hobble down the track, return with a monster Toblerone, devour the lot, and then share a huge evening meal. By the sixth day I was desperate, and my friend had kindly realised that it was not much fun for me to see him heading off with two girls every day to be left alone with *The Idiot*. So, as a surprise, he arranged for a taxi to take us all to a beach on the other side of the island, his grandmother packing us a fine picnic.

It was on this stony beach, unable to go into the sea, the plaster up to my knee slow-roasting my leg and dressed in minute black Speedos that I first felt uncomfortable in my own body. Whatever loss of weight had come from our ascetic trip through Europe had been more than offset by the last week. In the middle of the day we retreated into the shade of some tables and umbrellas nearby, and were joined by a Greek couple who looked as if they had been drawn by a cartoonist.

They were probably only in their mid-forties, ancient to us, well-heeled and local, dripping gold medallions. The wife had expensively executed blonde highlights over an incredible hourglass figure. The husband had thinning grey-black locks brilliantined back over his burnished head. He sported a mighty walrus moustache in jet black, and a deep brown body which had entirely gone to seed. He was introduced to us as Oscar.

They had known my friend since he was a baby, his Greek father's oldest friends, and though they had no English we sat happily in each other's company for a while. When they left us, my friend explained that years ago Oscar had been the most handsome man on Kalymnos, famous in Athens. His body was thought to be so perfect that his nickname was given him after the Academy Award statuette.

Still quite young, I found it impossible to picture this fattened-up Greek as a young, firm god, and said as much to my friends. They looked at me, as Peter Buckroyd had, surprised that I could not see any parallels with myself. Later that afternoon we stopped in the taxi at my shop and one of the girls generously bought one of the huge Toblerone bars which she meant to make an event of for our last night in Greece. Ashamed, I didn't take my fair share, leaving my friend's grandmother asking me if she could make me something else instead for pudding.

That summer of 1980 was drawing to a close. University would not start until October, so I returned to Harrods to squirrel away even more essential funds. I had been to a paradise and, rendered impotent by my feet, like Oedipus I'd made a mess of it. In that last short session of the year in Knightsbridge I returned to my peanut and milk diet with renewed vigour and continued to struggle, jogging on pained shins in the autumn air. For university I would be beautiful again, and those 28-inch black Levis would hang from my hips once more.

Twenty years later I looked across at Kalymnos from Turkey. I was sitting alone one afternoon with my friend's pretty wife, a fine doctor, as Kylie's 'Can't Get You Outta My Head' throbbed through the taverna speakers. She looked at me and gently asked if I minded her asking a personal question. This, I thought, is interesting. She opened her blue eyes wide and said, 'Is there a history of obesity in your family?'

Genoa

In those last months back at Harrods I was assigned to the
stationery warehouse pushing pallet trucks of posh notepaper
through the bowels of the store. After each trip to the shop floor I
rode my empty trolley like a skateboard through the long tunnel
under the Old Brompton Road to our dimly lit base. It was a huge
cage in acres of open warehouse. Waiting for me there among the
shelves of biros were a homesick Nigerian studying more than
working and a young man called Nigel who talked about his
mother a lot and dreamed of promotion to front of house. All of us
were overseen by a kind and elderly Greek lifer also called Paul.

The glamour had gone. Other year-offers were still in Phuket
or doing good deeds in the Sudan. I was losing my tan, pretending my foot didn't hurt, and cramming in the overtime. For
mental stimulation I was on my own, taking yet another volume
of Freud to Hyde Park and, if I was working a long Saturday,
slipping down one of those shiny yellow custard tarts in both the
morning and the afternoon breaks.

From the beginning of my time with Freud I always thought
that he was a good read rather than a great doctor, an example
sans pareil of the wise man driven to bullshit to earn money and
keep his family. That October I was still clutching his works

when I headed for Exeter St David's station for the first time on the Inter-City train from Paddington. The three-hour journey would have been enough to keep him busy for a week, because although I was fearless the oddity was overwhelming.

The entire train seemed to be packed with young people from private schools in Surrey, but I didn't realise that hurtling down another line from the north of England were young people from comprehensives in the former industrial heartlands. This mix gave Exeter University its character, and some genuine grit, with the likes of me incidentally cast as the social lubricant. The Northern Scientists were chalk to the Wellies' cheese. The former despised the élan with which the latter came on to campus in their VW Golfs, wearing green Wellington boots, Barbours, and sometimes leading a muddy Labrador into lectures. The NS feelings turned to rage when a hardcore cell of Wellies drove through Exeter in a white Rolls-Royce, throwing money in the air, while the Northern Scientists were striking for an increase in the student grant. To the Wellies the Northern Scientists didn't really exist, making the NS faction even angrier. It was unreformed Labour versus high Tory, and it felt as old as the hills.

The odd feeling that I was sending myself to boarding school when I was old enough to vote haunted that journey, but what Freud might have defined as my unconscious discontent was offset by the discovery of British Rail Genoa cake, a plastic-wrapped rectangle of densely fruited stickiness which was the railway's only triumph in the 1980s and happily remains so today. Communist Genoa had been my first stop in Italy earlier in the year and, though I doubted the Italians would have recognised this moist slab, the association gave me pleasure.

As the train drew into the station we could see the university above us. It was built as a campus on a number of steep hills to the north-west of a once beautiful city, which had been bombed to smithereens by the Germans and appallingly rebuilt. They left just enough pockets of the medieval near the cathedral and down at the quayside on the river Exe to give it some residual character.

I came to love it, and love it still, but that wasn't the moment for appreciating a flawed Devon city. Because the word on everyone's lips, as I unpacked my case in a ground-floor room I was to share with a fine lad called Geoff from Somerset, was Sex.

Exeter had just been nicknamed Sexeter by the *Sun*, which was responding with characteristic good taste to the very high figures for abortions and STDs quietly reported by the campus medical centre for the previous year. Meanwhile, the *Daily Mail* had reported that the runner-up to Miss Great Britain 1980 was about to arrive as a student and every man was quivering with anticipation. Then, when I attended my first Psychology lecture, I discovered that I was one of six men in a year of sixty.

The unscrupulous might have made hay in this environment but my romantic life was already tangled. I thought I was still going out with Sister Sarah back in London, though neither of us seemed really sure. I did attract the attention of some female students, but discovered quickly they were reading Psychology for a reason, and if I had wanted to pursue their interest it was going to involve hours of listening to sobbing about their parents' divorce first.

I decided that the best strategy was to step back for at least the first term. Astonishingly, I must have given off some sage aura of aloofness, which acted as a magnet to the sweet but surrounded Miss GB Runner-Up who was being pursued by every oafish rugger-bugger and arty poseur on the campus. My friendship with her gave me an early insight into the genuinely awful situation a beautiful woman finds herself in when she has been tainted by the juices of celebrity. Later that year I came close to the most notorious example of this in British history.

So what I did at Sexeter University was to spurn the opportunities available and form an ardent crush on the one Psychology student whose background was a bit like mine. The epic difficulty was that because she had the most extraordinarily pretty features every single man she ever dealt with was as drawn to the flame as me. She had more bother than Miss GB Runner-Up. So although there was a memorable evening in Tiffany's nightclub involving Dire Straits', 'Romeo and Juliet' ('You and me, babe, how 'bout it?' And then Mark Knopfler tingling both our spines with his Fender) it proved impossible to follow through, especially when for her, like me, there was another character off-stage.

An unfortunate pattern developed. During the week, everything was a thousand miles per hour. Romantic intrigue, study, midweek parties down by the quayside, living for the next moment of being in a room with her. Then, at weekends, she would usually go home, and so would Geoff, and I found myself with long and lonely Saturday afternoons in a Devon city which did not yet feel like home.

This was the first time I felt the dull rhythm of life in bedsit land. I could hear the distant crowd from Exeter City football ground half a mile away. The depressive sound of Joy Division's 'Love Will Tear Us Apart' leaked from every window in a hall of residence pokily partitioned from a Victorian house.

It was all so detached and weird, and sitting in front of two bars of a gas fire reading Freud on a weekend afternoon, fretting about the love affair that could never be, was not going to help. For Sigmund believed that the human psyche tends towards stasis, and that when everything is in balance (or half-dead) a state of ordinary comfort can be found. He lived by his own advice, and when he was evicted from his Viennese flat by the Nazis in 1938 and permitted after international pleas to be exiled to London it was the first time he had moved his belongings in fifty years.

I was about to discover my own deep seam of comfort. When most people hear the name of Kipling they probably remember the man who advised us to treat triumph and disaster as two impostors just the same. I do not, for it was on one of those late winter Saturday afternoons that I first found stasis by discovering a range of products which are mainly responsible for the man I am today.

Treacly Track

The greatest twentieth-century product launch in confectionery came from a standing start, the Mr Kipling cake brand. Although it seems as established now as Victorian names like Cadbury's it was the brainchild of the advertising agency J. Walter Thompson and only got going properly in the early 1970s. JWT were responding to the realisation by Rank Hovis McDougal that they had a surplus stock of flour, sugar and fats. A creative genius called Jeremy Bullmore was commissioned and he immersed himself in the history of the British and European Cake, thereby rediscovering and relaunching an array of manufactured treats we now take for granted which, before then, were fading into oblivion.

My friends used to say of me in the consumerist eighties that I would have been a useful freak to industry because I was entirely resistant to advertising. That was true, to the extent that I didn't Tell Sid, believe That'll Do Nicely or wish to slip on, even if I'd been able, Nick Kamen's 501s. But every time I heard the slogan 'Mr Kipling Bakes Exceedingly Good Cakes' I was as enfeebled as a mast-bound sailor hearing the sound of Sirens.

French Fancies, Bakewell Tart, Almond Slices, Viennese Whirls, all were carefully launched as 'exceedingly good cakes'.

In a class of its own was the product I stumbled upon during a Saturday spent wandering sadly to a corner shop in the Victorian back streets of Exeter. It was to be my alpha and my omega: Mr Kipling's Treacle Tart.

I carried this ravishing thing back to my lonely lodgings and switched on the radio for news of the latest mishap from Charlton Athletic. Then, in an act as absurd as the Mad Hatter's Tea Party, I took tea with myself. Waiting for the results announcer to say Forfar 4 East Fife 5, I took the tart from its foil tray and sliced myself a chunk, bleeding syrup from the cut. I sat back in one of the pair of old people's home armchairs, let out a sigh mingling contentment and resignation, sipped some tea, and slowly sucked my way through the first piece.

It was on one of those low teas, which I would date to about February 1981 (I remember now that awful weekend, Valentine's plans dashed when she'd left for home the night before), that I first abandoned all the rules and customs of civilised eating. In company, I would have shared the treacle tart happily and generously. Alone, one slice led to the next, and so on until my plate was empty. I didn't yet realise that my carbohydrates were empty too, but even I knew that by eating an entire treacle tart I was loading enough energy to power an elephant.

The best thing to have done in the circumstances was to skip supper but, impoverished as I was, I did not feel I could afford to. Our first-year digs were all in, and two of the meals of the day were followed by hot puddings with custard. I was the hall cook's most appreciative customer, particularly fond of her sponge topped with jam and desiccated coconut, her spotted dick and above all, strumming my Achilles heel, her syrup sponge with custard which with Roman decadence came with a jug of warmed syrup on the table.

The next best thing would have been to go for a jog. But by the evening the halls of residence seemed to be alive again with young men and women wanting to go to the pub and, although I never really drank like a student, I was beginning to look as if I did, so in the bar of the Edgerley Hotel I washed away my feast and my troubles with three half pints of Skol.

By that first summer and aided by the false testimony of a girl who said, 'She'll never really go out with you, you know', Exeter became less of a source of pain and more of a home. That June I reached a fork in the road. I hadn't been on the scales since I'd arrived but, with the quiet accolade of having once been London's second fastest Under-16 runner over 400 metres, when a whippet of a girl asked me to join the Athletics Club, I reasoned that this was just the thing to blow away my residual blues.

In my first race meeting on the Clifton Hill track with its westerly views to Dartmoor I pulled a hamstring. In my second I won but vomited for half an hour afterwards. My third race which, as they say, came too soon, was at the University Athletic Union Annual Championships at Crystal Palace.

I had caught the train from Exeter that morning, been at a party the night before, and was incredibly anxious that both my ex-girlfriend from London and my father (who was obliged to stay loyal to Betty's misconceived view that she was 'bad news') were in the crowd. It took me an hour to find my team and get hold of the right vest and number, leaving about five minutes to warm up in the pouring rain.

I stepped on to the track and for the first time I realised that I was absurdly ill-prepared, out of my depth. A grown-up but disillusioning thought came to me in the cold light of this rainy day – just get yourself round the track. It had never been in my character to be so pathetically resigned, and it sickened me, but I knew it was of course what I had to do.

My strategy changed, however, when the tannoy announced the names and teams of the runners who'd be racing against me.

'Lane 1 Someone from Somewhere. Lane 2 P. Arnott, University of Exeter. Lane 3 Someone Else from Somewhere Else. Lane 4 S. Coe, Loughborough College.'

I looked up from my spikes towards the man standing two lanes outside me. He had a sleek Concorde nose, legs to just beneath his rib-cage, the tensile strength of a ballet star and the confidence of an Olympic Gold Medal-winner, which, in Moscow the year before, he had become. Though the 800 and 1500 metres were his real events Sebastian Coe was using this meeting as an early season sharpener for greater things later on. He was the greyhound, I was a rabbit.

Quietly I seethed. He was a good five or six years older than me. What kind of student was that? What was the degree, Seb? Metalwork? The resigned calm of a few moments before was completely shattered. My insides were ash, my mind hot and my forehead covered in sweat.

Before I could begin to reason what to do, orders came to climb out of the trench, the starter fired his gun and we were off. From this moment a fundamental difference between the future Lord Coe and me became clear. He was a middle-distance runner using the 400 metres to get faster. I was a 100 metres runner who'd always used the 400 metres to slow down.

Panicking, I applied the starting strategy from the 100 metres, and went off like a rocket. Within a few explosive strides I'd caught and passed him on the inside. All I had to do then was get through the next 360 metres.

It was at this point that the paucity of my preparation began to tell. There were shot-putters in better physical condition than me for this race, and every organ in my body except my troubled brain was still little more than stone cold. As we approached the first bend and the spurt of adrenalin which had got me there was spent, I began to feel the tangible tearing of the hamstring I had pulled earlier in the year. Millimetre by millimetre it was steadily ripping. Pride made me want to go on, but reason told me I had to pull up.

I looked to my right and saw a huge green crash mat in front of the main stand on which pole vaulters fell to earth. It was inviting me to clasp the back of my left leg, swerve across the track and throw myself on the sympathy of the crowd as I winced bravely, attended by the St John's Ambulance Brigade. That is what I did.

What I had not allowed for was the crash mat being covered in about an inch of chilly rainwater which had been gathering all afternoon. As I threw my wounded carcass upon it I landed on my back but instead of stopping in an elegant pile I began to aqua-plane across it, picking up considerable momentum until I skated, arms flailing, feet first from the other end into a freshly raked sandpit. That hurt.

S. Coe, Loughborough won the race, taking two seconds off the fastest 400 metre time ever run at the University Athletics Union Championship. I returned to Exeter by train that evening, my sore leg twitching in my seat, leaving a girl behind I had hardly been able to speak to. The best thing to do was keep the leg moving, so I walked along to the next carriage, and then the one after that, as far as I could go, until quite unconsciously I found I'd entered the buffet car, where jovial West Country lads back from a day at the races were spilling beer and cigarette ash on the floor.

I returned to my seat where I took the lid off a cup of tea, and unwrapped a slice of Genoa cake, wondering gratefully how they got a huge cherry into every single slice.

Tea at the Palace

While better-heeled students went off in the summer of my first year to enlighten the Bushmen of the Kalahari or water ski the Aegean it was my lot to replenish my empty Nat West bank account by another stint at Harrods. For a week or so I was back in its Estate Office, a nineteen-year-old selling houses in Montpelier village to Arabs who draped their dining rooms like Bedouin tents.

This was a job for my pinstripe suit. The previous year it had fitted where it touched, now it fitted like a glove. My Harrods photo ID had changed too. Last year's was all cheekbones and eyes. The new one had a glazed expression, the hint of fat pads diminishing the eyes in a head adorned with the bigger hair of the early 1980s.

During this first week I used my discount card to buy a genre of book which was relatively new to the market, certainly to me, a slim and alarming volume giving the calorific value of every common food. As I sat at my desk waiting for the phone to ring from Kuwait I blanched at the realisation of what I had been shovelling down at Exeter. More than anything, I was shocked when the maths revealed that my patented peanut and milk diet would have put blubber on a walrus.

Many pompous documents have been headed with the title What Is To Be Done, but on a piece of A4 card designed to take the particulars of applicants looking for a flat in the Cadogans I began to map a revised dietary plan. Next term I would be living in a student house with three girls. There would be a need for resolve on many fronts. My understanding was that I would have to build my life around the jacket potato.

Before I could memorise the number of calories in cottage cheese my phone rang calling me to see the Head of Personnel. Urgently. As I was standing to leave my desk a Harrods messenger arrived bearing identical orders on headed paper. What on earth had I done wrong? Had the day I'd clocked in for a friend the year before come to light in a belated audit?

The Head of Personnel was the kind of woman who Terry-Thomas would have described as foxy. Plainly she had something of huge importance to tell me but she let me sweat a bit first in her hot office with its view of the drainpipes in a central stairwell. Suddenly, she changed gear, and the tone of the meeting went from low-level harassment of a male employee to a scene from John le Carré.

It was essential that I listened to what she said. A very important establishment had contacted Harrods to ask if they could provide them with a trusted employee for a special task over the summer.

I remained impassive.

The necessary checks had been made through the secret service and I would be pleased to know that I had been approved.

My impassivity act slipped.

'This letter is for you. I know its contents but it is sealed. You need to give me an answer in the next five minutes.'

I left her office and sat in one of the hundred or so seats used at recruitment days in the empty personnel department. The letter bore the crest of Buckingham Palace. It offered me a job and a salary double what I was then being paid. It told me where to report and when to do it.

Mobile phones were not available for another decade, so I went to a call box and phoned some friends. My instinct was to refuse. I really didn't like the extraordinary messing about with my civil liberties, the arrogant presumption that I would drop everything to go and work at the Palace. Plus this proposed servitude required me to work my whole vacation and miss out on a £69 all-in coach holiday to Lloret del Mar I had pencilled in for the end of September.

I did, though, very much like the sound of the money. I had a minute left to decide and everyone's phones just rang and rang.

The job title on my contract was Assistant to the Chief Accountant and Paymaster. It had been created as a temporary role because on 29 July that year Prince Charles was due to marry Diana Spencer. The Palace decided to recruit through Harrods because mine wasn't the kind of job you filled from the filing cabinets at Alfred Marks.

Because of the wedding hundreds of extra staff were being hired, from construction to cleaning. All of them were paid in cash. My job was to sit behind a huge desk like Bob Cratchit with an immense ledger and calculate with a fountain pen how much each of this legion was due. I then took the appropriate notes from the most enormous pile of cash and folded it into manila envelopes.

When I wasn't doing the payroll I was bringing security bags bulging with even more cash from the Queen's Gallery to the side of the Palace. Only the knowledge that this was the most fortified building in Britain and a dash more of that institutional arrogance could have allowed so much money to be so casually handled.

In theory I was also meant to be preparing a series of Rolodex cards about each permanent employee in preparation for the Royal Household computerising its operations at some unspecified future date. The staff I met in accounts could hardly wire a plug – an abacus was still in operation – and although by now this system must surely be up and running I wouldn't bet the house on it.

The staff with whom I worked were faceless, loyal, unambitious and woefully paid. Their wages were microscopic but that was to miss the point. This was old-fashioned service, duty, recognised by Her Majesty herself with a personally presented surprise gift and a lick from her favourite corgi at Christmas. Ultimately, for anyone who did their time in the name of their sovereign, it resulted in a minor honour and an even tinier pension, but principally the rich privilege of the kind of memories my summer there gave me.

There was one substantial perk for which Her Majesty's staff did not have to wait. At 3.45 p.m. every afternoon the pens and ledgers were set aside and we all made our obsequious way along red-carpeted, mirrored corridors to one of the Queen's dining rooms. Once there, we were required to take afternoon tea.

This was not mentioned in my job description, and after a couple of days I asked my boss if I was obliged to go. He left me in no doubt that even to frame the question was an insult to the sovereign and by implication a slur on my fellow worker bees.

I simply couldn't tell him that this was like forcing an alcoholic in recovery to go to the pub every day, that the last thing in the world I needed, spending sedentary days behind a royal desk, was to be served a huge tea, identical in specification to that served to the most important Ambassador to the Court of St James.

At first I didn't tell my friends what I was scoffing every afternoon. I took it very seriously that I had signed the Official Secrets Act. If the Russians heard the royal household was bingeing on salmon sandwiches, scones, and Fortnum & Mason cake instead of micro-managing the royal wedding where would this lead?

Even more compromising would have been the information that on our return from the dining room all we did before clocking off at five was reshuffle the papers we'd left on our desks and metaphorically pick our noses. But soon I had to share this greedy secret. I was already living too many paradoxes.

I was skint but working cheek by jowl with absurd wealth. I was a closet republican but giving my summer to the greatest royal publicity exercise ever. And I felt incredibly idle, gorging cake with flunkies, when every afternoon in the next office a young woman a few months my senior was working her way, without a break, through sackfuls of mail which wished her and her future Prince the best of British for their wedding.

I doubt it figured on the radar of MI5, but on many evenings that June I'd sit like an apprentice Falstaff with my friends in the Black Horse trying to give a picture of the pantomime playing out around this twenty-year-old woman. She was of an identical background to many young women I knew at Exeter but without the A Levels to protect her. She had strikingly blue eyes but when I spoke to her in the corridor she seemed to be living, over an agonisingly prolonged period, what I'd suffered for just a few moments at the legs of Seb Coe – the self-knowledge of being in too deep.

I'm not sure how or why I was invited to the wedding. Someone must have put a word in, but I was called into my boss's office and he said he was surprised to learn that an invitation was to be issued, albeit for a pew half a mile from the altar. This was a step too far for a republican so like an idiot I said that I couldn't go and please would he decline on my behalf straightaway. High tea felt like high treason that day, and that evening a gang of us Bardolphs made plans to take a 'Get Away From the Royal Wedding' day trip by ferry to France.

Once there, we kept the Union flag flying by getting impossibly drunk, running through the streets of Calais with our bottoms out, and playing a game of soccer with dockers by the quayside. Finally we heaved our innards into La Manche on the way home.

We couldn't get away from the wedding after all. Every deck had a television. I went back into work the next day to pay even more wages than usual. Naturally enough I felt unwell.

The next day I was struggling too. It was a welcome feeling in one way, as if my excess weight were burning off me with every minute that passed. By the time this feeling had gone I had lost three stone.

No Fat

I have only managed to lose weight three times, each as the result of serious illness. Pneumonia and an emergency appendectomy both proved good for losing about a stone, but the infective hepatitis I contracted that summer of '81 was the real deal.

The symptoms were interesting. The first was profound exhaustion, to the point of being unable to shuffle even a single inch. I put this down to my treasonous excesses in Calais as I worked what would prove to be my last day at Buckingham Palace before going to a sickbed for the rest of the summer.

No matter what offence I would cause to the Royal Household I simply couldn't make it up the stairs for a slice of Dundee cake, and my five-minute walk to Victoria that night took half an hour. It was hyper-strange, especially as the two people who'd caused the Palace to be in a relative frenzy were now far away and set on their historic course.

The second fascinating symptom of hepatitis, if this is your area, is a change in the stool. It becomes innocuous and babyish, because all the toxins it would normally be carrying from the body are now swilling about in the bloodstream. After this came a phenomenon like a fragment of science fiction, a change in the appearance so dramatic that friends actually hesitated before they came in to the room to visit me. I went entirely yellow, and not the benign hue of a bowl of good custard but a dayglo effect halfway to green.

This is unappealing enough, but it's what it does to the eyes which frightens the horses. My orbs were naturally quite large with a bluey centre, but where my whites had once been was now a violent green. I looked like Paul Atreides from the *Dune* novels I adored in my teens. As a youngster I envied his immortal life sustained by a mysterious blend of extra-terrestrial desert spices. I had not expected to end up with his eyes.

How did I contract this unpleasant disease? On my shortlist were a girlfriend recently returned from the Land of the Pharaohs, the dodgy moussaka we shared in a dodgier restaurant during a rude night in Earls Court, or just about any unwashed activity of the previous few months. It might even have been a Buckingham Palace gateau. This lurking infection had incubated in my liver until such time as I got thoroughly pissed, which rare event occurred during my confused trip to Calais.

There is no medicine for infective hepatitis other than the passing of time. The most important part of the recovery is to immediately eliminate any form of animal fats from the diet as these are what the liver is having most trouble processing. You also have to give up alcohol. The liver has a great capacity to mend itself but if you drink while it's recovering you will add scar tissue or cirrhosis.

If you are calorifically minded you might already have calculated that this non-fat, non-alcohol regime is one of the surest recipes for fast and dramatic weight loss. When I returned to Exeter belatedly for my second year I weighed in at less than I had at sixteen when I'd sprinted from one end of school to the other for the sake of it. And for three months I had not eaten a sweet, biscuit, cake, chocolate bar or milkshake for the most sustained period in my conscious life. An opportunity to break an ingrained pattern had emerged, but what would I do with it?

Attempting to help a brilliant young director launch a production of *Class Enemy* by Nigel Williams in the dualistic atmosphere of the student theatre society – half-Rattigan, half-Brecht – I accidentally won the role of Iron, a Brixton class bully and hard man en route to a breakdown. At last, the time had come for me to dress like Marlon, tight red t-shirt, leather jacket, and falling from my shoulders to my elbows ridiculous biceps developed through the thousands of press-ups I had been doing since my illness in lieu of sex.

The production did a couple of little tours, and within a few years four of us had gained the compulsory wings of the theatre, in the dying days of the closed shop, an Equity card. Looking at my photos now, I look less like Marlon and more like Posh Spice, with my stick-thin body, big bonce and inflated individual parts. But it was cathartic, a season playing an angry young man, and after the premature middle age of my over-responsible late school days, a taste of youth.

By my own standards, my consumption of cake and chocolate during this period was marked by a Buddhist self-control. Well beyond a time when it would have been safe for me to take up fatty stuff and alcohol again I sipped a gnat's piss called Barbican alcohol-free lager and nibbled only rarely at a small bar of plain Bournville chocolate.

I remember this time as one of achievement and vibrant isolation. Everything in my former character said it could not last. It did not. Soon I was finding my unerring way back into the comfort zone.

Deflation

If I were sitting by a fireside now with my two sons and they asked me about the birds and the bees I'd think carefully. Best not say anything too dogmatic, always be available as a father and a friend who'll judge neither them, nor their choice of partners. Bite my tongue, and bite it hard, because the only thing I ever learned was to be careful about going out with actresses. The odd thing is I always knew this deep down, and the curiouser thing still is that I met every single one of my four girlfriends between the ages of seventeen and twenty-six when I was acting with them. There were compensations, but there was drama too. Turbulent, bordering on the insane.

As my second year at university went on I found that the role of lone wolf in a leather jacket didn't really suit me. It was by definition lonely. I must have looked as if I needed a girlfriend, because the next thing I knew I had one. The difficulty with the undergraduate life is that it's always cheaper and cosier to move in with someone, and this is precisely what happened to us. From rediscovered youth back to middle age in under a week.

My new girlfriend was lovely. From Surrey, of course. But she had rather a blunt way with words. I would often stay awake at night as she snoozed wondering what she'd meant by 'I'm happiest on a horse', 'You're just my major undergraduate romance', or by a casual comparison with the one before me.

It was a pity that we hurtled towards domesticity so quickly. Soon, haricot beans were being soaked overnight for nourishing stews the next evening. Avocado, mozzarella cheese and granary bread became compulsory. This was heading for disaster, not least because we were eating this stuff like the Waltons at night, just in time for us to pass out and turn the whole lot to flab.

The seriousness of the situation became clear when I was invited home to meet her parents, who were charming but also incredibly frank with each other. Still as vain as a peacock, I imagined I was going to go down well. In addition to my new domesticity I was President of this and Chairman of that, a fine figure of a man and, I reasoned, quite a catch. Indeed, when I'd been snaffled I felt there were a number of disappointed rods still left dangling in the water.

The first lunch at the family home in Surrey was a bit of a tussle. My girlfriend's father was Chairman of the Bar Council, and had the most sensitive bullshit meter I have ever encountered. For a twenty-year-old waffle merchant like me he was a tricky adversary, but we seemed to rub along.

Her mother seemed more appreciative of my obvious promise, and as I stood next to her by the sink, drying up, I felt she was won over. Later in the afternoon we swam as she watched us from the kitchen window, and in my trunks I felt like Burt Lancaster swimming his way across American pools in the sunshine.

That night, I lay naked and resplendent waiting for my girlfriend to return from the bathroom. She slipped out of her dressing gown and huddled up next to me.

'I think Daddy likes you,' she said, which was good, because I liked him, and if he hadn't liked me breakfast would have been awkward.

'My sister thinks you're funny,' she continued.

'Ha-ha, or weird?'

'She didn't say.'

This unsolicited, unprocessed feedback slightly pissed me off. Still, I must have scored a hole in one with the mother, and sought the inevitable compliment.

'Mummy says you're charming.'

My head swelled. There was more.

'But that you could run to fat later on.'

My head and all other parts deflated.

'What the hell is that supposed to mean?'

'Don't get scratchy.'

I got out of the single bed and strode around the room.

'Me. Run to fat?'

'She's older than us. She's seen a lot more life than we have.'

A Wildean aperçu about never mind me, what about her enormous arse, wisely remained unstated.

'Yeah, well. She's wrong. That is just plain bollocks.'

'All right, all right. Come back to bed.'

I did, and continued to do so for the rest of my time at university. But I never did like Reigate.

Mars

The problem with telling the unvarnished truth to people you are fond of, even love, is that they are much more likely to listen to the same message from someone else. It was one of the group I'd bonded with in *Class Enemy* who first grasped the nettle. I was playing Belville in *The Rover* by Aphra Behn, a Restoration comedy written by Britain's first female playwright, full of fops and tavern wenches. I was the boring male lead, decked out in puffy sleeves, knickerbockers and a tatty wig.

My fine friend came backstage after a dress rehearsal and said simply, 'You look like a fat poof in that costume.'

A shrewd fellow, he waited for me to change into my normal clothes before developing his theme. I'd put on a favourite blue jumper with yellow stripes, and made my way up the hill with him to the student bar. He was still looking at me with his beady eye, and concealed his follow-up remark behind the name of his cherished girlfriend.

'Maggie says some people have to be very careful wearing horizontal stripes.'

We stopped walking. I said that I didn't understand.

'Well, you know, it's obvious. You're the psychologist. Perception. Visual cues.'

We walked on some more. I thought I saw the picture.

'What you're saying, Nick, is that you and Maggie don't have to worry about horizontal stripes. But I do?'

'Yeah, well, you're a muscly bloke, aren't you?'

'And that the reason is if you present the human field of vision with a series of descending parallel lines, and the ones in the middle are more stretched than the ones at the top, then people will perceive that the subject causing the distortion is in fact overweight.'

'According to Maggie, yes.'

While Nick ordered me a half of Barbican I went into the loo and looked at myself in the full-length mirror. What was he on about? I was bloody gorgeous. During our friendship I'd never mentioned the fact he was only five foot six, but perhaps it was time for a few jokes about wearing lifts in shoes.

Then as I walked away I caught sight of myself in profile, and the entire theory of the horizontal stripe, twenty years before Trinny and Susannah, lay bare before me. From the side I saw that my yellow stripes were starting to droop at the front.

Back in the Ram bar Nick was just about to pay for my drink. I saw that with characteristic generosity he had bought me a packet of Worcester Sauce Wheat Crunchies to go with it.

'Are those for me?' I said.

'Yeah.'

'Please, put them back.'

It was a sad moment, as if I were recognising I had some kind of problem. College life had begun to lose its simple pattern of the year before, and one complication seemed to be piling upon the next. The easy egotism of acting was replaced by chores of producing and directing, as well as devising and managing a programme of theatre companies visiting the campus. And at the end of every day I had a relationship to sustain, and the smaller matter of a degree to study for.

My various responsibilities earned me half an office, where I sat sorting contracts for companies I'd spotted in Edinburgh or submitting applications hoping to chisel a hundred pounds to underwrite their visit from the local arts authority. My only break was to go downstairs into the Devonshire House student café, where I'd have liked to stay and listen to the gossip from the latest party on the moors. A courageous friend raised money from the gilded youths who were the most striking habitués of the café, and with it made a short film of an F. Scott Fitzgerald story, 'Bernice Bobs Her Hair'. He was trying to make a point, but as ever it was taken as a compliment.

My own escape was not the Super8 camera but the chocolate fudge cake served in the sandwich bar. At 50p a slice it was not cheap and, still living on a cheese-paring budget, I often had it in lieu of a proper lunch. When the pleasure of that wore off I would go downstairs again to a little shop and buy a Mars bar.

As I sat staring out of my office window at the settlement of Exwick in the distance, and the hills beyond beginning the incline to Dartmoor, I nibbled the edges around my Mars bar until all its nougat was exposed. Holding it most delicately, I then bit away the nougat until what remained was just the stiff chocolate top and the toffee layer sticking underneath.

Then, it was in with the front teeth to chew and lick the toffee away, and on red letter days I found myself with just the long, bare layer of chocolate, which I nibbled away like a squirrel finishing his favourite almond. How much more enlightened would my pleasure have been if I had known the story of Forrest Mars who developed the eponymous pride of American confectionery in England, using British engineering genius to make layers of chocolate, nougat and toffee adhere without melting into each other.

I would love to have pondered on Mars's terrible falling out with his confectioner father in Chicago, his self-imposed exile in Europe. How he secretly infiltrated the factories of both Nestlé and Toblerone in Switzerland as an ordinary Joe, and then located his new empire in Slough because the rents in central London were too high. But I didn't know any of that. At that stage in my life a Mars bar was not there to interest my conscious mind, it was there for pleasure.

The paradox with the raw delight it gave me was that the Psychology I was studying was meant to teach us how to look for the science of the human mind. Freud and all who followed him were constantly theorising around the very word, pleasure. Was it more Thanatos than Eros, more concerned with death than life? Was there some kind of oral fixation involved?

Looking again at his original writing, especially as, in the name of cake, I became familiar with his home turf in Vienna, it is brilliantly specific to its time and place. He had a superb literary but only moderate medical mind. He was hugely ambitious but wickedly thwarted for promotion because he was Jewish. Because of this Jewishness he had an ancient culture, civilisation and sense of family bonds to draw on, much richer than the local Austro-Hungarian Christianity.

Having convinced himself and others that free association as they reclined on his couch would eventually disclose underlying problems, he then made a living from it. Observing that process being repeated thousands of times, he recognised similarities between patients and teased out some structures of the mind. He defined them as sexually based. He would have run a good writers' course. Read today his division of the stages of formation of the psyche into oral, anal, phallic and genital seems almost parodic.

If I'd have been on his couch at No.19 Berggasse and told him about my interest in the sweet stuff he'd have had me down as orally fixated, and made a diagnosis about the relationship between this and either being gullible or sarcastic or both or neither, based on the linguistic pun that the oral person either swallows everything they are told or conversely has a biting tongue.

That I was not breastfed because I was adopted may have won me my own monograph, but the tragic truth for Freud is that his stuff is as reliable as astrology but twice as interesting. One dear

Psychology professor at Exeter was trying to prove via statistical analysis the mathematical existence of the Oedipus Complex. It had been his life's work, and it was a credit to the British university system that he was allowed the time to do this. But in clinging on to ideas nearly a century old, he showed that if there was such a thing as the anal phase he was still in it.

One of Freud's great inspirations was Sophocles' *Oedipus Rex*. I read it carefully, but the only character I found interesting, especially as he was not bound to live out a tragic path, was the blind soothsayer Tiresias. One of his lines hit home. *To have perception where the perceiver draws no profit is a dreadful thing.*

What I clearly perceived was that between the platefuls of chocolate fudge cake and the regular flow of Mars bars I was tucking away a mountain of cocoa and sugar. The question that was preying on my mind was, in modern psychological terms, was I developing some kind of addiction?

We were fortunate that one of our best lecturers was an internationally recognised figure in addiction studies. He was six foot tall, tanned, a surfer at weekends. A strong-jawed Geordie of Viking stock, he looked as if his entire life was macro-biotic. It was a privilege to study with him as he explained the behaviour underlying the abuse of alcohol and drugs. One afternoon when the lecture theatre emptied I stayed behind and we talked.

As we chatted I gradually got round to the question I wanted to ask.

'Is it possible,' I said, 'to have an addiction to chocolate?'

Even as I said it I started to laugh, and he certainly did when he realised what I was asking. He gave me a short, improvised tutorial I will never forget.

He explained how in addiction studies a little knowledge is a dangerous thing. That because we were beginning to understand a tiny bit of brain chemistry we were in danger of bandying words like seratonin around as if its manufacture in some spurious pleasure centres in the brain was the be-all and end-all of human life.

He said that studies showing that the psychological effects of eating chocolate were akin to those experienced during sexual intercourse epitomised the worst theoretical practice of coming up with a sexy hypothesis and then bending the evidence around it. He said that anyone who has actually seen a poor soul craving alcohol or morphine, going cold turkey, would not dare to use the word addiction anywhere near the word chocolate.

As he sat next to me across a table he cleared this small worry of mine away. He had also predicted the next twenty years of bullshit about craving, longing, needing, desiring, living for and being addicted to sugar, cake, sweets, biscuits and chocolate.

Finally he looked concerned for a moment.

'You're not diabetic, are you?'

I didn't know what one of them was, but said I wasn't anyway.

'Are you sure?'

I said I was sure.

'Well, that's all right then. You're just greedy. Who isn't?'

Battenberg

It is well-established by historians of the British royal family that Lord Mountbatten of Burma held great sway over the lives of both the Queen's husband, his nephew Prince Philip, and her son, Prince Charles. Many insist it was Mountbatten who advised Charles to remain a happy bachelor for as long as he could until, with his wild oats sown, duty obliged him to marry an inexperienced young woman for the purpose of propagation.

One didn't have to dance in a pill-box hat outside Greenham Common airbase to find this an unacceptable view of the world. That the girl selected was the one I saw innocently opening Good Luck cards in the corridors of Buckingham Palace, as if she were about to take her driving test, left me unimpressed by Mountbatten and all his works.

Many historians have thrown graver accusations. That he bungled the partition of India, that he had a secret army prepared to run the Britain of my childhood under martial law if the trade unions carried on striking and that there were reasons he was known in the Royal Navy as Mountbottom.

Of his family's one act of greatness, however, there is no doubt. When his ancestor married one of Queen Victoria's granddaughters in 1884, a cake was named in her honour. She was the Princess Victoria Hesse-Darmstadt, and she was marrying Prince Louis of Battenberg.

For those concerned with *Burke's Peerage*, Battenberg became the name of a morganatic cadet branch of the grand ducal family of Hesse without right of succession. If they and all the inbreds ruling Europe had sat down for a cup of tea and a slice of Battenberg the First World War might have been halted in its tracks.

Failing this, the Battenberg family anglicised their Germanic surname as fast as they could – to Mountbatten. The legacy of this was twofold. They would not be blamed for the war and were accepted into the British ruling class. But by abandoning their good name they could never take credit by association for a cake which ranks with Golden Syrup as one of the great sweet treats on earth. For the wily Lord Mountbatten himself the change was perhaps as well. Answering to the name of Lord Battenberg would have carried as much gravitas in the nation's affairs as if he were called Lord Walnut Whip.

At about this time in the early eighties the playwright David Hare wrote and directed a film called *Wetherby* set partially during post-World War Two rationing. A young woman, played by Joely Richardson, first meets her boyfriend's family over a stiff afternoon tea. Wanting to compliment her prospective mother-in-law, she asks her how she'd made her lovely Battenberg cake.

'Don't be stupid,' cuts in the boyfriend, 'you can't *make* Battenberg.'

That was always precisely the point. If my mother, Betty, had her friends round for tea the cake stand was mainly loaded with what she'd baked herself.

But fanning out sluttishly on the bottom tier, out of the way so as not to spoil the impression of home-made meringues and fairy cakes, were four or five provocative slices of Battenberg. Her friends would say with complete conviction, 'I couldn't manage another thing, Betty. I'll burst. No, no,' and then polish off the lot with the dregs of their tea just before they got their coats.

There are recipes for Battenberg (and mine is on p. 221), but that misses the essential idea that it ought to be shop-bought. That is the British way. During my time at Exeter the intellectual furnace that is the Mr Kipling Product Development Unit gave the Battenberg an unimaginably fresh twist. Understanding it was impossible to improve on quadrants of pink and yellow sponge wrapped in marzipan, they took the fundamental idea (sponge and marzipan), retained the Battenberg name, reduced the portions to pygmy size and dipped both ends in chocolate. If I were a dog I'd be dribbling.

Then, they presented it not as a whole to be sliced at home, but as individual cakes in boxes of six. Genius is an overused word, perhaps. Nonetheless, the Battenberg Treat was genius.

Precedent for my abandonment to gluttony had come with the Toblerones of Kalymnos, and then the treacle tarts of my lonely first-year weekends. For both of them there could be some plea in mitigation, whether because of physical injury or plain misery. There was no excuse for how I treated a box of Battenberg Treats.

Off to a corner shop, back to my place, strip the cellophane, rip the box, and take the first one down in one. Then two more. Put the kettle on. A mug of tea. Another two follow swiftly in an oral paste with the PG Tips. Then only one left. In it goes. Fulfilment. No need for an evening meal, listen to some music, consider the sad fading away of my spiritual life, off to see my girlfriend, burn calories away. A lifestyle to match any I have ever known.

At this time, when I was not grinding out some arty project or eating new confectionery products like a gummy old dear, I was whipping up an entry for a short play competition. Wanting vaguely to do something Freudy, I'd lined up a brilliant Jewish friend, later alumni of RADA and the RSC, to play the man himself. A natural entertainer, it also occurred to me that I could do something about the mental illness of James Joyce's daughter, Lucia. Running out of time, I put Freud and Joyce in the same play and won the competition. It was the kind of beginner's luck only a youngster who hadn't heard that Tom Stoppard had already done Joyce in *Travesties* could have.

The prize was £50. It came in the nick of time, because I was desperate with worry that I was now overdrawn at the bank. In 1983, going over by £5 attracted a stinging letter for which you were charged a further five. Somehow, though, it was agreed by the cast that the right thing to do was to go out to the most expensive restaurant in Devon and blow the lot, subsidising this feast by a further £10 each. It was impossible to explain to such a lovely group of young people who had no concept of financial desperation how much on that occasion I could not enjoy stuffing myself silly.

That first play led to another, properly produced at the Northcott Theatre in Exeter. This was the time of my third actress, a clear example of what was happening backstage being a thousand times more dramatic than what was happening under the lights. This play also had a psychological bent, and tried to encapsulate theatrically the difference between the psychoanalytic, behaviourist and cognitive schools in modern thinking. The musical rights remain available.

It is often said that you should write about what you know and, unable to write a speech about the third actress, who was already beyond words, I gave one character a zippy monologue about the Gestalt school in psychology. This should have been the most boring scene in theatrical history, but it was not. Since Gestalt was concerned with how the human mind was greater than the sum of its parts, examining a whole entity by reference to its components but not defined only by them, the audience ought to have been asleep in the aisles. That they were not was because the best analogy I could conceive of to explain Gestalt was the Battenberg Treat.

Although it seems far-fetched it had the advantage of being the truth. If you dissect a Battenberg Treat into its constituent parts you're left with sponge made from white flour, a vanilla-flavour paste smeared around it, and marzipan with a very low almond content. The whole entity dipped at either end in a chocolate substance about as low on cocoa solids as consumer law would allow. Spread out on a plate like a body on the slab, it seemed an unattractive combination of third-rate individual bits of food chemistry.

Take another one from the box, however, in its livery of yellow and brown, and it's beautiful again, at least to the eye of a like-minded beholder. Bite it, even knowing what it is made of, and the way it yields, its small burst of sweetness are deeply satisfying.

After the show a lot of people came up and spoke to me about 'that Battenberg speech'. Roughly they were in two camps. Those who thought I'd gone mad, and those who too seemed to understand what I was on about.

For although we were in the era of the New Romantics with a long stretch of Thatcherism ahead of us, I was not the only one who was failing to engage in either project. In an individualistic age of unbridled competition some of us were naïvely hoping that we might yet be able to have our cake and eat it too.

Pusillanimous

The times had changed us all. I graduated and made my way
back to London where, in a deregulated world, youngsters like
me were launching enterprises which once would have caused
men twice our age to pause for thought.

A few decades earlier those who wanted to put on plays for a
living spent years being chased round their desk by producers like
Binkie Beaumont. In a depressed economy where you could hire a
theatre and put a show on yourself there was no such apprentice-
ship to learn your trade. You had a go, and if you crashed you
burned.

I was partnered in a new production company by an Old
Etonian. He was incredibly funny, very good-looking and, as he
hadn't lobbied his way into Pop at school, a young man of
modesty and ambition. We went about our business putting on
worthy shows about subjects like Steve Biko, where my partner
had the flair to persuade Peter Gabriel to allow use of his Biko
music to go with it. We did a musical too. These projects didn't
make us money but just about paid their way, and took us into
the penalty area looking to score.

Somehow we managed to persuade John Osborne's agent to
permit us the rights to the first ever West End revival of *Look*

Back In Anger. I was twenty-two, and can only assume that we talked a very good game. We found a small venue, the Arts Theatre off Leicester Square, and put down a deposit. We'd be ready in July, a year after we graduated, for our first hit.

The crash was tragic. My partner's girlfriend was still at Exeter and, taking herself off on a spring walk near Lyme Regis, she fell down a cliff and lay undiscovered for three days. Her head injuries were serious and her full recovery would take years rather than months. My partner's place was by her side and so we postponed the summer production until November. The Arts Theatre agreed to this and held on to our deposit. By the time we were ready to ink the contract for *Look Back In Anger* another young man in a hurry had snaffled the rights instead. He was Kenneth Branagh.

That November he opened all-guns blazing to great reviews at the beginning of the Ken & Em phenomenon. Stuck with our booked theatre, we put on a new play. Our reviews were less kind than Ken & Em's. We limped on and closed a fortnight early. We did not have a last night because the lighting gantry went up in flames.

The whole endeavour was so foolish I can only see the comedy in it now, but at the time it did not feel funny. Seeing the way the wind was blowing, and broke, I took a job back at Harrods. I was driving a van delivering hampers to the Home Counties by day, and appearing in this crucifying play by night.

Whatever I was paid seemed to slip through my fingers trying to keep the show afloat. I was then going out with my fourth and final actress. My lowest point came the night I realised that her cat, Billy, was eating better than me. I had enough pennies for a bowl of rice with a squirt of tomato purée, while Billy feasted on Marks & Spencer's sliced ham. Let it never be said that actresses don't know how to look after a cat.

My partner and I were left to survey the wreckage. The debts were awful, and some went unpaid. Nearly two years later I had saved enough money to take the actress on a cheap holiday. That was the day we received a court demand from our own lawyers. Another miserable London summer ensued.

The saddest aspect of it all was the breaking of a beautiful friendship. We had shared a flat for a year or so with the easy domesticity of Morecambe and Wise. Seeing me at breakfast once, he wondered if I realised that if I didn't do something about it I was going to end up looking like Willie Whitelaw. He didn't mean to be hurtful, but weight was an issue for him too, in a way I had never experienced.

His legs were like pipe-cleaners (with the apparent compensation that this made his penis look huge) and, having spent time in Africa, he didn't eat puddings on principle. Urged by my own parents to finish my plate in honour of the starving children of Biafra, this was a position I never understood. The paradox was that he knew he was a natural customer for the Charles Atlas flexing machine. Unlike me, he was actually desperate to put on weight.

My advice was to stop smoking forty Marlboro a day and, having made his tribute to Africa, start eating puddings again. When I saw him last he did not seem to have aged a minute, patrician and lean with a fag dangling between his fingers. I looked like Willie Whitelaw.

Back then, however, there was a parting of the ways. We went off to make our separate livings, but in the mid-eighties this was not as easy as it might appear. Unemployment was rising fast, and if you had a job you had to hold on tight. At least I had my Harrods van. But whatever weight I'd gained during the torment of the failed show was about to disappear. Eating a jam sandwich in my freezing cab the following February by Putney Bridge in the coldest winter on record, my weakened frame began to succumb to pneumonia. When I recovered, the job and the actress were no longer mine.

Greek

If you've had hepatitis then pneumonia is a stroll. You can eat what you like but you can't really be bothered. If you're lucky and live in the western world pneumonia is treated with antibiotics and then by a physiotherapist who pummels your lower back so powerfully that anything lurking in the lungs comes rasping back up your throat. When it's over you have lost a satisfactory amount of weight and are not left yellow from top to toe.

When I emerged blinking into the real world it was the same only worse. I wasn't depressed, but my circumstances were. I was a roadie for the Silk Cut Dominoes Tournament for a while. These were deeply sedentary events held in places like Glasgow where men and women would play dominoes all day, plied with free cigarettes by local models in purple sashes.

It would be illegal now. It should have been then. At every location the winner was presented with their prize by an ex-footballer. I realised that if the first British man to hold the European Cup, the ex-Celtic skipper Billie McNeil, could stoop to this, then I might have further to fall.

The quiet streets of not-yet-on-the-up Highbury Hill saved me in a way. They gave me a home in north London for the first time. I'm not sure how I would have coped with such a come-down in my old haunts south of the river. More importantly, the nearest corner shop was simply the worst tip I had ever seen. The milk was sour, the bread was mouldy, and the Mr Kipling cakes I might easily have given into in my reduced circumstances were never less than a month past their sell-by date.

This was a barrier I have never and won't cross, though the evidence shows I am too timorous. Later that same year I helped a friend move into his first house in Wimbledon. In the garden we dug up a dead cat, and behind a kitchen cupboard we found an intact box of Almond Slices date stamped to eat six years earlier. We opened them out of curiosity, expecting a green sludge. They were immaculate. Later that weekend my friend rang to say he'd just found pictures of the former owner's wife in the back of an old wardrobe with, in his own words, a cherry up her arse. About as enticing as the Bakewell Tarts in that dirty old shop.

Back in Highbury there was nowhere to go to eat anything at all. Every other Saturday a man set up his burger stall in the scrub of our front garden, but with our bird's-eye view of his health procedures when he slapped a burger into refried onions and lard, we were never tempted to eat there. The bugs he introduced into football fans' guts every other week probably went unno-ticed, as they were all half-plastered from their pit-stop at the Plimsoll Arms in St Thomas's Road on the way to the ground.

On the rare occasions when income exceeded expenditure I chose to leave my surroundings and take the Piccadilly Line from Arsenal tube station right opposite my flat. I was always mildly excited that it had me at Leicester Square in under fifteen minutes.

Like a fruity old thesp I bought a *Stage* from the newsstand and took it into Gaby's Delicatessen and Diner next to the Wyndhams Theatre. In here amongst other displaced persons I could buy a feast for under £3.

I began with warmed falafel and hummus in pitta bread,

chocker with generous piles of fresh salad I wasn't getting anywhere else. As I read of the early career of some friend in rep, I relished this meal as much as any I had eaten, sipping slowly, to make it last, at a Diet Coke.

With the falafel gone I was nearly full, but with just sufficient room I went back to the counter and pointed at the precise slice of shamali I wanted. I didn't want a corner, and I didn't want a dry end from the day before. I wanted a central slice of this Greek almond cake which was absolutely drenched with sugary syrup around its base, just about holding up a fine, golden sponge and a single sliced almond. The syrup was infused with rosewater too, so this seemed like both a Greek and a Turkish delight.

Shamali is the only cake I have never eaten two of at a time. I am not sure even I could manage it, so pure and strong is its sweet syrup. But it is also so plainly a food of the gods that I have too much respect to bolt two at once. It would be like eating a second lobster, over-indulging to diminishing returns. I associate shamali with recovery, giving it an almost totemic importance.

Those feuding cultures of the Eastern Mediterranean, the Greek and the Turkish, share so much in common in their cuisine that it is hard for strangers to tell them apart. The Greeks, though, are the masters of sugar. To them it would be abhorrent to eat falafel and then follow it with shamali. Greek sweet baking is meant to be eaten at a café for its own sake, a meal in its own right. It used to be made mainly with my old enemy, honey, but thankfully there is a lot of refined syrup in use now.

We are becoming used to baklava in Britain, layers of filo pastry drenched with honey or syrup. We also know kadaifi, or Angel Hair, the one that looks like a Shredded Wheat immersed in sugary fluid. Harder to find is the stunning galaktoboureko where layers of sweet-soaked filo are sandwiched together with firm vanilla custard.

What we haven't developed yet is the habit of offering friends who come for tea not a biscuit but glyka tou kautaliou. This translates as Sweets of the Spoon. Your host takes the lid from a jar in which they have preserved what we would recognise as candied fruits, only they seem to be candied all over again and then stored in syrup. These can be apricots, cherries or peaches, but the sugar content is so high that they might even be aubergines or courgettes.

The guests dip in their spoons and, as they catch up with the gossip, they lean over a plate and nibble the delicious sugared fruit. This beats a Fondant Fancy any day. It is the old men, though, sitting outside harbourside cafés, watching their sons tinker with boat engines, who seem to have the best idea. Just as my children like to eat sugar lumps as they come, so these men eat ypourichio, or Under The Water. This is simply sugar boiled with water and vanilla essence until it becomes a thick white icing, which they eat from a teaspoon kept moist by dipping it into a glass of water and then giving it a good long suck.

Food worriers today will warn of diabetes and obesity and other terrors from eating in this way. But these men are ancient by the Aegean. Because they are happy. My next job was playing the happiest old man of all time.

Custard Tart

There are some jobs you are chosen for which it is best just to say yes to and not ask why. The phone rang as I sat in my flat counting coppers to buy some teabags and a charming lady called Cheryl introduced herself. She told me that she worked with Michael Bond, author of the *Paddington* books, and that I had been recommended to her for an important job they had starting soon at his London HQ.

This sounded interesting. I hadn't considered children's fiction, but a writer must master many crafts. Cheryl asked if I would mind coming in for an interview at the Paddington Bear shop just off the Edgware Road. She explained that it was the main worldwide outlet for all bear-related products and that she and Mr Bond had an office upstairs.

We agreed that I'd go in the next morning, and just before she ended the call I asked if she'd mind telling me something about the job, so that I could think about it overnight.

'Oh,' she said, 'hasn't Richard told you? I'm so sorry, I assumed he'd phoned you.'

Richard was my girlfriend's father, a rather good musical actor as underemployed as thousands of others.

'Yes, he immediately thought of you. We offered it to him

first, of course, but I think he's got something down at the Orange Tree for Christmas so your name came up.'

I tried to think fast. What job that might be offered to a fifty-year-old actor would also be suitable for me? I couldn't think of anything.

'Yes, he said you'd be perfect, because you've got such a jolly face. We've still got the costume from last year. Richard did it for us then, absolutely terrific. If you're able to do your own beard and make-up we'll be happy to reimburse any costs.'

The phone in the flat was near a mirror. I looked at my reflection. The penny had dropped. I was twenty-four years old, yet a highly experienced actor who'd played the lead at the London Palladium had been asked to suggest someone for the Paddington Bear shop gig, and the first person he'd thought of was me. Six weeks, thirty-five pounds a day, cash in hand. Playing Father Christmas.

The next morning I wasn't sure what to wear. I wanted the job because I needed the money. I hoped to seem young, reliable and keen. I had my interview suit but thought that'd look too *Wall Street* for Santa. My worry was irrelevant. The interview was the only occasion I have ever had to dress up in a costume for a job.

I felt like a young girl in the hands of a glamour photographer when Cheryl asked me to go behind a curtain and slip into the red trousers, the red coat and a huge leather belt. And the hat with the bell.

There was no mirror, and no beard, so I had no idea how I looked. I pulled back the curtain and did my best to walk towards my interviewer in a Santaesque way.

'Marvellous, Richard was spot on. A very jolly face. Now, what do we think about padding? I don't think we need to, do you? Just try altering your stance.'

I stuck out my stomach and my pressed my hands into the small of my back.

'Great. And you'll be sitting down anyway. Good. I think cushions look so ridiculous, don't you?'

For a moment I wondered if Kenneth Branagh was playing Santa this season. Then I accepted the job. I walked out on to the Edgware Road knowing I would spend that deep midwinter listening to the hopes and fears of children who trusted I had magical power. I suddenly took it very seriously.

But as I passed the Lebanese shops with their sheep's testicles and honey-soaked cakes something even more lasting than common Christmas cheer came into my life. A hundred yards from the tube station was a large branch of Marks & Spencer and, as I looked up at its green lettering, a chain of thought began. The Santa job meant money. So I could spend what was in my pocket. Which meant that I didn't have to eat spaghetti, a chopped tomato and dried basil again for supper. I could go into the M&S Food Hall.

If ever a leopard changed its spots in British retail it was M&S in the mid-eighties. A few years earlier it epitomised British paralysis, grey clothes for a grey country. Then seemingly overnight it had bought some refrigeration units, shoved the beige slacks to the back of the shop, and presented the world with a range of foods we would wonder how we ever did without.

There were marvels galore. A neapolitan bar with milk, plain and white striped chocolate. A delicious corruption of the Walnut Whip, filled with coffee cream instead of vanilla, or with no walnut and a plain chocolate button instead. The reintroduction of the coconut mushroom. Rhubarb and custard sweets a master sugar baker would praise. Marzipan bars covered in chocolate.

Vanilla slices in the chilled zone, and mini sherry trifles. Pots of individual chocolate sundaes with a pool of dark chocolate fluid in the last spoonful at the base. That November afternoon is when I found one of the great foods of my life: M&S custard tarts.

It was tricky enough that they were sold in boxes of two. But they were sold in boxes of four as well, for just 89p. Suddenly the London I so despaired of wore a smile. I felt as rich as any Arab on the Edgware Road as I carried my M&S bag full of custard tarts on to the best-looking stretch of the underground, the high vaults of the Circle Line heading to King's Cross. On to the Piccadilly Line and ten minutes later I was buying an *Evening Standard* from the man at Arsenal tube and stepping the few yards into my flat. I was alone, but not really. I was alone with four custard tarts.

How much I revelled in them it is hard to relate. Nutmeg, vanilla, mace. Cold eggy custard. Butter pastry. Silver foil cups in a yellow box, with transparent windows, so you see them naked waiting for you to come in. I ate all four and between them I can't have drawn more than a dozen breaths. Christmas had come early for Santa.

Yule Log

I was sharing a flat with a half-Asian actor who was so talented at their school in Salisbury that Ralph Fiennes credits him with being his first inspiration. Sadly his ethnicity still hampered him then, so he was doing a worthy Shakespeare season in Tufnell Park. He had trained at drama school and advised me on how to make a beard from a pair of tights, some glue, and coils of white hair from a theatrical costumier. My girlfriend chipped in too, and before long I looked like Sir Donald Wolfit playing King Lear in a Japanese black and white film. I was ready for my throne.

My days of playing Father Christmas followed a daily rhythm. Tube, Paddington shop, costume, Ho-ho-ho, Marks & Spencer's custard tarts, home. And the more the weeks followed this rhythm the more jolly I looked. Was it the inner joy of pleasing so many innocently bewitched children? Or was it the tarts? Maybe it was both.

I learned many strange things sitting on a gilt chair in my heavily themed Paddington Bear grotto. The first is that Arab children will come to see you with their bodyguards standing in the tinselled doorway and, although they don't speak English and haven't got a clue who you are or what they are doing there, the universal benevolence of the Santa archetype gets you both through.

The second is that you don't have to be scared when a fully-grown American sits on your toadstool and says he's been praying to you but you haven't been answering him any more. Just behave like a therapist. 'I've been speaking, Bob, but maybe you haven't been listening in the right way.'

The third is never ever use Santa as a threat. So many mothers ruined my day and that of their child by saying, 'Father Christmas won't come if you're a naughty boy. Will you, Santa?' And the fourth is the reason you should never say the third. There is nothing more open than a child gazing into Father Christmas's eyes and telling him the simple truth. The hope and love in that one look will reaffirm the faith in human nature for the most cynical of Santas.

I have never been so at ease in a job. It even rubbed off on some of the mothers and one or two of my elves, from whom I felt an extremely festive vibe even from behind my whiskers, or perhaps because of my whiskers. And it made me wonder what it is about the shape of Father Christmas that is so adored. He cannot ever be played by morbid skinnies like Jeremy Irons. He must be well covered.

Maybe it is because he has fallen. He has experienced life and grown fat on it but this has also made him a whole man. He is not a tormented teacher who wanted to be in the circus. He is not a parent who has not slept properly for five years. He won't shout at you, indeed he is the one adult who never will. He is a free, fat, funny little man who brings you things you want and spreads unconditional love from the back of his sleigh.

There was one elf who I used to joke with during the day and before long we knew each other's life stories. She had never seen me without my costume.

On the last day before Christmas we were incredibly sad because we knew we'd never see each other again. As I pulled off my beard in front of her she saw me for who I really was and told me that I'd make a lovely old man. That remains the greatest compliment I have ever been paid.

I went off into the cold December night feeling like a layer of skin had been removed. Young again in this hostile city. Still, there was Marks & Spencer across the way, and there was no need to allow some small insight into the human condition to stand in the way of another box of custard tarts. And maybe a pot or two of brandy butter for the spoon.

Beating my usual way towards the chilled cabinet as the shop prepared to close on Christmas Eve I saw the many cakes and mince pies still stacked to the ceiling. All I wanted was the tarts really, but food photography is an art of its own and I simply couldn't walk past the Yule Logs. When I got one home I opened it immediately and cut a generous end slice. It was horrible, the ganache far too rich and fatty, the outer chocolate not honest Cadbury's but something pretending to be Belgian. A few moments later I had tipped the whole thing into the kitchen bin, and to make sure I didn't relapse I emptied that into the dustbin out on Highbury Hill.

So, I thought, I'm not just a big fat greedy old Santa. I still have some taste. I'm not just anybody's. Luckily I had stocked up with custard tarts too. I was all theirs.

Fat Hamlet

There's an old saying everyone who's ever crafted a review wishes had been left unsaid. If you can, do, if you can't, teach, and if you can't do either, be a critic. This is cynical. In my young idealistic way I felt nothing but a sense of responsibility when in the year after my season as Santa I began to cover fringe theatre first nights for the new *Independent* newspaper. Its listings pages packed with reviews of just about everything became a crucible for many senior journalists of the future whose cvs to that point were as incoherent as mine.

There probably are critics who lead bitter, vicarious lives, fumbling inside their trousers by day and dipping their quills in poison by night. Most don't. When you are young it feels like a privilege to give your published response to the best efforts of others, one that simply shouldn't be abused. Since I had a painful background in the business side of theatre I knew the guts it took to put on a show, so I tried to be scrupulous in what I wrote. One good write-up for a fringe company could secure an Arts Council grant for five years. One careless review for a play above a pub about the life of Wordsworth might give the impression it was another young writer's play about a literary great. Yet for the right punter it could shift a perspective for ever.

The *Independent* really was independent, and while the rest of the British media became infested with the eighties disease of being horizontally and vertically integrated, taken over in over-leveraged deals by dodgy new proprietors, it stood bravely alone. Those fortunate to write for it, even in as tiddly a role as mine, were appreciative of this and carried on refusing free trips and taking tiny payments.

When I wrote my debut review I was still so poor that I didn't have a bank account and had to cash the cheque through someone else's. It was a very rare evening, however, when I left the flat thinking what I was about to see was going to be irredeemable. On dozens of the hundreds of nights I had in London and eventually across the country I came back and communicated to my flatmate what I had just seen with sincere astonishment and some zeal. That the very oddest of these shows was so near to my favourite subject of cake seems weirdly predestined.

The show had an eye-catching title, *The Pornography of Performance*, and was a loosely structured performance-art piece from Australia. It packed them in at the Riverside Studios in Hammersmith possibly because the actors were naked. They were also prepared to stand in full-height cylinders with holes cut out, allowing audience members to grope blindly up and down their bodies. On the night I was there this seemed to have attracted a fair constituency of single men with comb-over hair carrying plastic tartan shopping bags.

I gave it a measured review. In truth it was a bit sixties and ran the risk of making Australian art seem provincial, but it was sincerely done and as long as you weren't going there in lieu of a proper human relationship it was okay on its own terms.

This review got me involved in a contender for the worst TV programme ever transmitted in the United Kingdom. It was called *Club X* and was the forerunner to *The Word*. It strove to break new ground on Channel Four but usually fell backwards into any hole it had just dug.

Club X was broadcast live late-night from an ill-equipped warehouse near the huge gay pub south of Vauxhall Cross, the Vauxhall Tavern. The studio floor was populated with heroin-chic skinnies and the gangstas of tomorrow. It was chaos. It wasn't creative chaos. Nobody had a clue what they were doing.

The company of *The Pornography of Performance* was invited to play out one of their scenes live. I was cast as the fair-minded, youthful critic. Teddy Taylor, the hard-right Tory MP for Southend, was to be Mary Whitehouse. One of Europe's leading entertainment lawyers was going to comment on censorship issues.

For a cartoon right-winger, Teddy Taylor was, before transmission, a civilised and entertaining man. Indeed, he was so mild-mannered I wondered what he would really have to say. The nudity of the show was surely no worse than strip pubs in his own constituency. I had underestimated the *Club* X producers.

They had chosen a part of the show which, when performed on a distant stage, had some theatrical force, but when performed six feet from the end of the nose of Teddy Taylor MP was probably the lewdest act ever broadcast in Britain. This extract involved two totally naked male performers walking slowly towards us, the nervous panel.

One of them was then upended by the other, so that he was standing on his head. Having achieved perfect balance upside down, his colleague then opened his legs wide so that he made the shape of a catapult. Things started to look hairy.

A tray of cream cakes was then brought on solemnly by one of the female performers. It seems incredible that I can't remember whether she was naked or not, but what was about to happen simply wiped the image of her away. Pausing only to look around the studio and eyeball Teddy, the upright Australian then took the largest of these cream cakes and began to ram it into the back passage of the upended fellow. He did not cease until he had appeared to insert the entire trayful. There was cake everywhere, and when the man sporting the creamy innards came back to the upright there was more.

The studio lighting changed. As if to preserve their modesty the performers were cast into the shadows, though they were still naked, while a harsh spotlight was cast on to the three wise men. The presenter, an insincere young Irishman, predictably invited Teddy to respond first. You could ask Robert De Niro to depict 'shocked' a thousand times and he'd never conjure an expression like Teddy's.

Teddy was not, however, struck dumb. His tirade was un-stinting. He pointed to the creamed actor and spoke out in his Heart of Midlothian accent.

'Oh, my word. I feel so sorry for you. What a disgusting way to have to make a living. You have just been abused for the sordid entertainment of the witless.'

This was not what the presenter expected.

'So you're saying it should be banned?'

'No, not at all. Is that why you have dragged me into this place? I do not believe in censorship. You clearly have not done your research. I just think what happened to that young man just now is terribly sad.'

I hugely enjoyed Teddy's response. It was incandescent, genuinely shocked, and showed a surprising empathy for the cast. Instinctively I felt that there was a bubble of manipulated outrage growing in the room, partly blown up by the actors but cynically inflated by the TV producers. The Irishman turned to me for the considered view of the *Independent*.

'From a personal point of view,' I said, 'it's a terrible waste of cake.'

I wasn't going to give this stunt any credibility. I waffled on a bit longer, there was a bit more solemnity from Teddy, and then the Irishman responded to something in his ear and cut as fast as he could to some pre-recorded idiocy on tape.

The funniest moment of the night was the expression on the entertainment lawyer's face when he realised that this part of the programme was over and he had not had the opportunity to speak.

'Is that it?' he raged. 'Are you not coming to me?'

The Irishman blathered about running out of time.

'You have not run out of time. You have plainly cut this item halfway through. I trust you will be coming back to me after this tape.'

The Irishman explained that was that as far as *The Pornography of Performance* was concerned.

'You mean I have been in this dreadful place since early this evening and you will not allow me to contribute my views. My children are watching this at home. I have a serious legal practice.'

He stood up and like all amateurs who lose the plot on television found that he was still tangled up in sound cables.

'How do you think this makes me look? I've been seen to watch a man have cream cake inserted into him in front of my face and then don't appear to have anything to say about it afterwards? On national television.'

All he did not say was that *Club X* would be hearing from his lawyers.

Thus as a young and temporary critic I interacted with the wider world.

Sometimes, though, even as an onlooker, I was in a position to give encouragement which was sincerely appreciated. There was in my estimation nothing especially brave about garlanding your buttocks with pastries. The production I really admired at that time was called *Fat Hamlet*. It took its inspiration with tongue in cheek from the tragic hero's opening soliloquy: 'O! that this too, too solid flesh would melt.' It then winkled out all sorts of other references to weighty matters which the audience had never been aware of before.

Playing Fat Hamlet was a truly Fat Actor. This took guts. What you understood in his unusual portrayal was that he didn't get along with his mum because her vanity made her ashamed of her son's looks. Moreover, when this lump of Danish bacon tried to quietly debate a point with one of his relations his sheer bulk terrified them, even when all his intentions were meek. It was a powerfully original production, and at a time when the issue of multi-racial casting was to the fore it was a reminder that multi-size casting might be an issue too.

Fat Hamlet himself wrote a letter of thanks after the review was published. He was pleased that someone somewhere understood what he was trying to do. This is one of the rarest and most precious experiences in any professional life. As I dunked a dozen Digestive biscuits into a mug of tea and worded my reply I didn't mention just how empathetic I was.

Big Bang

Everyone who has ever dreamed of writing but needs to earn an income too cites the example of T.S. Eliot. Famously he was a poet before breakfast but a London banker for the rest of the day. It's too grand to claim that what I did next was proper banking but, following his example, I was an aspiring radio writer before breakfast, missing funds officer on an hourly rate by day, and theatre reviewer at night.

I was employed at the heart of the City of London precisely at the time when it became deregulated. The generation before me had found a way to generate banking, accounting, legal and management fees by selling the railways, power companies and telephone exchanges my parents had paid for since the war. We never thought they'd sell the water too but they did.

Behind all the spivvery lay the long shadow of the traditional City of London. I was working in an investigation unit trying to trace missing international transactions often involving millions of pounds. It was clerical back-office work. My fellow workers were the real pulse of the City, men and women who joined the bank after the war and were approaching retirement. Old fellows like Mr Mallard, defeated by the photocopier, who asked, 'I wonder if you might help me photograph this document, Mr Arnott.'

These City folk would not have dreamed of burning their wages in oyster and champagne bars. They had been going to the same pub off Cheapside for forty years. They may have lived in Croydon but it was they who marshalled the Lord Mayor's Show, who were sidesmen at the dozens of historically unique churches within the City walls. They made sure these churches hosted thunderous organ concerts at lunchtime, when it often felt like Bach himself had just played for us. While in sharper-suited offices elsewhere the yobs were selling the family silver, they were all that was good and sustaining in the City of London.

However, there was one custom of their long service which did not have a good effect on a man in my condition. The concept of the luncheon voucher was as ingrained in their terms and conditions of work as the provision of a helmet for a policeman. Many of the younger traditional City workers would eat a meal at lunchtime paid for entirely with vouchers and by eating nothing but a few slices of toast at dinner time they saved up the deposit for their first homes. That was how many young marriages began.

I would have loved to have been in one of those couples meeting at Cannon Street for the train back to Bexleyheath, but I was still madly involved with a dramatic actress. The chances of us ever eating together in the evening, or at least being sure of doing so, were remote. Occasionally a knife might have been drawn to dice a carrot but it was usually whizzing past my head before both ends had been chopped off.

Which is why I went with my luncheon vouchers and these gentlemen of the City to their favoured, subterranean, barn-sized restaurant, the Long Room, which looked like it was still blacked out from the Blitz. I would take my tray to the long counter by the wood-panelling and lavish my vouchers on a good square meal for lunch.

The difficulty for a young man who didn't feel overweight but kept getting little hints that he was on the slide was that the restaurant's idea of a square meal was a plate of roast meat and vegetables one might eat at a pub halfway through a twenty-mile walk.

Moreover, while more experienced men took an apple and did the *Telegraph* crossword at the table, I ran my eye longingly over the huge catering trays of crumbles, sponges and puddings. For those who joined me at the counter it was plainly a reminder of the school lunches I never had, and was the main reason it was packed with men enjoying good stodge from their earlier simpler lives.

With these mighty desserts came huge jugs of toppings. Custard, chocolate custard, strawberry-jam sauce, runny syrup and, at Christmas, great buckets of brandy sauce which I could have drunk all afternoon. The ladles you used to help yourself were bowl-sized, and while some took a disciplined quarter measure I filled my ladle to the brim and poured it over my bowl until it was ready to burst its banks.

On a bright day I might walk some of this off, trying to navigate my way through to the banks of the Thames, which the Corporation of London had carelessly allowed to be blocked off with concrete over the years. I wandered along Milk Street, Pudding Lane and Sugar Quay but despite such evocative names the City seemed cauterised from its colourful past. Everyone else was shopping.

For culture I read in the Barbican library or went to one of the church concerts and was transported by the musical ecstasies of long-dead composers from Mittel Europe. Soon most of the blood from my brain drained towards my stomach to digest a mound of spotted dick and custard and I would doze off like the other bone-tired men and women, our jaws hanging open. We awoke when the organist pulled out all the stops for a grand crescendo. There would be polite applause which turned to a grateful tumult when the musician took a bow from the organ loft.

For most of us this was the only expression of emotion in a long working day spent trying to find out why a transaction begun at the Bank of Benghazi through the Union Bank of Switzerland had never arrived at the Banco di Nicaragua. On paper, the chain of money might inspire a Freddie Forsyth novel, but as I struggled for consciousness in Gresham Street, dictating telexes to shifty characters across the globe, it made for one dull afternoon after another.

In those pre-CCTV days I'd slip from the front of the bank at around four on the pretext of going to the loo and buy a Mars bar from the tobacconist's next door. This, like my sherbet lemons at the Nook fifteen years earlier, had absorbed some of the odour from the other shelves. As I chewed into the soft nougat it reminded me of the Clan pipe tobacco an uncle used to smoke. He'd been one of these loyal aldermen of the City too and had the most impressive scars on both legs where he'd taken bullets on D-Day.

At the end of one of these long days I would think about the money and join the City migration by train, a more swanky journey now from Moorgate to West Hampstead where I was sharing a flat with the actress and three of her friends. I'd get out of the suit and head to one of a nonsensical range of productions, from Budgie – The Musical to a gruelling new Edward Bond at the Bush.

Finding myself alone and palely loitering one hot afternoon in West Hampstead I decided to take a walk towards the Heath. I put on some shorts and wondered about a shirt but with the temperature pushing 90° decided to go topless. The air was hot and wonderful when I stepped outside and I began to jog. Soon I had zoomed along Mill Lane, crossed the Finchley Road and was penetrating the urban village of Hampstead itself. The further I went the easier it seemed, and when there was a slope, rather than slowing, I accelerated even faster.

When I got to the Heath I felt hot but still full of energy, so I ran towards the Temple Fortune end and around Hampstead Garden Suburb. I had no idea my legs still had this kind of distance in them. I relished the perspiration coming from every pore and the hot tan I was developing with each stride. Still I went on, retracing my route and finally only ran out of puff as I came close to West Hampstead again.

I felt great and my natural assumption was that I looked pretty good too. But beware of seeing yourself as others see you, for as I walked across West End Lane I came towards a mirror shop on the other side. I was shining with sweat, but the mirror was in the shade so this made me look as grey as the mirror itself. As I came towards the corner with Mill Lane where the shop stood I saw that to the side of my lower ribs were two small flanks which in this unforgiving light were actually wobbling.

I could not believe it. I looped back across the road by a pelican crossing and pretended I needed to have a look in an estate agent's window down the way. Then I crossed the road again, slower this time, to check if this wobbling image had been a true likeness. It had.

Like Adam, I felt stark naked, and ran away up Mill Lane towards our flat as fast as I could go. I showered quickly, then dressed in the cleanest, most-pressed clothes I could find. I checked myself in the mirror and everything seemed fine again. That was the last time I walked the streets of London or anywhere else without a shirt.

That night I shared the experience with the actress. I was looking for kindness, I suppose. Her response was that one of her actor chums had seen us together on the tube and asked who the fat bloke was. I said, probably rightly, that she shouldn't trust a syllable from that particular, conniving, talentless, shitty, undermining, little broken excuse for a performer, whose every word was directed towards entering other people's pants. She added with customary lack of concern for appearing egotistic.

'I don't like it. It reflects badly on me.'

She had a point, but the next day in the Long Room I had the bread and butter pudding anyway. And when I got home from work, while she rehearsed *The Crucible*, I went to the corner shop and bought a box of Lyon's Cup Cakes in orange, lemon and chocolate flavours, although to be on the safe side I ate the icing and chucked the sponge from the last four into the bin.

New Berry Fruits

The era of the actresses had reached the final curtain. During the Great Storm of '87 I spent the night alone in my top-floor flat in Blackheath waiting for the roof to blow off. After an almighty battering without so much as a Rich Tea biscuit for solace, I survived and walked up on to the heath in the morning and across to Greenwich Park. A school friend, who existed in the basement with his smelly cat, and I saw the uprooted trees and smashed cars and felt the intangible feeling of emerging into a new age. The change to the landscape was devastating but the human toll was low.

Typically for a newly single male I was fit again. I burned the accretions of north London away on a running track in Sutcliffe Park, and was confident enough to remove my shirt in low-lit interiors, if not outdoors. The girl who had driven me to treacle tart at Exeter reappeared and seemed to appreciate my efforts, although one evening out of nowhere I felt a curious shot across the bows. 'My nightmare,' she said, 'would be to end up with a man with a pot belly, receding hair and a double chin.'

It was the most peculiar thing that, although much of the angst she inflicted on me seven years before was still within her compass, it was what she said about me with her professional hat on that brought me up short. She worked in market research, not quantitative, which tells you how many things you're selling and to whom, but qualitative, recording the reaction of research panels across the country to new or relaunched products.

There was no need for her to leave my flat to do that, I offered, since most of her research was in the launch of new chocolate bars, cakes and other items of confectionery. She was being highly paid to test-drive the new mint KitKat and I was able to sagely correct her on the history of Rowntree's prized product.

However, one night in a Blackheath wine bar I stopped talking about what I knew and listened to her for a change, to what she actually did, which was developing the psychological strategies for selling sweets by pinning down people's deepest attitudes towards confectionery. Over half an hour or so she mapped out the spectrum of products, as perceived by the manufacturers, and which part of the spectrum appealed to which demographic of the population.

That was the wilting evening when I found out my true position in life. My friend explained that it was possible to delineate preferred confectionery items by many variables, but one of the most certain of these was sex. For example, a real man eats Yorkies, KitKat, Mars and solid bars of plain chocolate.

That's interesting, I said. What if there were someone who did like Mars every now and then, but mainly enjoyed soft centres? Fry's Chocolate Creams, Turkish Delight, Walnut Whips, Crème Eggs. Anything with a hard outside and an inside you could nibble into, maybe suck a bit. I raised this in a blasé way, as if I was discussing a friend. What profile would she give to a person such as this?

She considered the science and replied confidently that we were definitely talking about an elderly lady, or perhaps a lonely younger female, from a mid to low socio-economic group with low status at a credit-checking agency.

I felt my balls clench under the table.

'Do you have a particular person in mind?' she asked. 'Is it your granny?'

I replied that sadly my grandparents were no longer with us. Just hypothetical.

Our future was clear. I might not grow into a pot-bellied man with receding hair and a double chin but I'd always have the dark secret that I was not really a man at all. I was Thora Hird. Did this mean I was the only man alive who enjoyed Meltis New Berry Fruits at Christmas?

Tarte Citron

I've tried hard not to break too many promises to my wife but I broke two on our wedding day. The first concerned her vow never to marry a man in a rented morning suit. It was just bad luck the old tailor stitching mine had a stroke the week we were due to be married, that no matter how I thumped on his old curiosity shop door in Covent Garden nobody ever came. She raised an eyebrow walking up the aisle towards me as I shimmered in nylon Moss Bros but was decent enough to keep on coming, despite my broken pledge.

The second promise I made a year earlier when we first met. She made a comment that I appeared only to own one pair of trousers. That's where you're wrong, I said. I have four more pairs at my flat. It's just that they are 32-inch waists and for some reason the 34s are fitting better this summer. I promised that within a month she'd see me sporting the 32s.

I never quite made it. In fact, I married her in a pair of Moss Bros 35s with all the vents let out. I am not yet in elasticated trousers, grounds for divorce, but when I see them for £15.99 a pair in the back of a Sunday supplement I fear they can't be avoided for ever. Many men will stand like Munch's 'Scream' at the 40-inch-waist rack, refusing to countenance the inevitable truth. It all seems so unfair. If women can change body shape over the years, then why can't we? My hips would bear lovely children now, if I was blessed with the appropriate internal organs.

What all young women who have planned their wedding for twenty years should know is that we grooms are just as neurotic about how we look on the big day. We feel paraded, utterly on display, and if you're not an inner catwalk queen this is very uncomfortable. I shared this precious confidence with my best man the night before. He used my coy confession to open up his speech.

'Before I start,' he said, 'Paul wants you to know his hair today is by Eugene at Trevor Sorbie and cost £39.'

He got his laugh, and as that was £34 more than I had previously paid for a cut it was not a cheap one. But despite the feeling familiar to all grooms that you are obliged to play the schizophrenic roles of both supreme romantic lead and butt of low wit in the course of the same afternoon, nothing could spoil my wedding day. It was still all of a piece with the way we met and the whirlwind romance that ensued.

The dateline was Normandy. We were both in Cherbourg covering an odd work of performance art which was having its première before coming to London later in the year. My finances had remained so tight for half a decade that this was my first trip beyond the borders in all that time. We were meant to come back on a ferry with some other hacks but we missed it and headed south to the town of Coutances.

The first days were all about retreating from the hot sun into cool incense-filled churches, and lazy hours sipping Stella and smoking like chimneys in bars. But they became about an extraordinarily Proustian moment too, when we both knew we were laying down a memory for life.

We had just walked through a hilltop park above the town and came down into a market square in the quiet hour just after the siesta. We'd skipped lunch and when we saw a pâtisserie so

French all it lacked was Depardieu twisting the croissants we were drawn wordlessly towards it. The sun was still high in the sky, but the yellowest light to be seen was reflecting from the window display of *tartes citron*. We claimed one each, an Orangina for our thirsts and a tiny and intense *chocolat chaud* for energy.

We sat under a parasol outside and sank our forks into the tarts. Our eyes met. Heaven knows what kind of lemons they had found, which eggs, how they had made this pastry, but it tasted ecstatic. We lived in the moment. We knew what we had found and ate slowly over a full half-hour until the heat had made the chilled lemon filling run across the plate and the pastry was warm to the touch.

When we had finished I felt I had finally found my mate, a young woman who felt as deeply about cakes as me. Of course, I hadn't realised that really she appreciated more adult tastes in food too, mussels, crab, venison, grown-up stuff I'd not encountered. But she knew a tart when she saw one.

The next weekend I met her friends in Regent's Park and we played softball. Seeing her hit the ball a long way and run like a human being rather than flicking her feet out sideways like a big Jessie I realised that she had passed all necessary genetic tests. My children would be in good hands.

A few months later I proposed on the Mount of Beatitudes overlooking the Sea of Galilee and we headed to Jerusalem later that day. I managed to get an unromantic dose of loose innards by foolishly buying some sweetmeat from a Palestinian's open tray in the Old City. At the American Colony Hotel, though, we had another pudding which made us sigh with our eyes wide. It was some kind of poached pear in juice, served warm, but it had a scent which seemed Biblical. We sniffed it and supped it for ages but could not work out what it was, and years later we realised it was probably cardamom. At the time it seemed an elixir, a tiding of good things ahead, benediction by dessert.

As the months went by I felt like the luckiest man alive. In love, of course, but also going out with a woman who was half-American. So what? Here's what. She wasn't from the Mr Kipling line like me. She made her blueberry muffins from a Betty Crocker mix. She made American pancakes on Saturday mornings. She made apple pie and crumble but with magical pinches of cinnamon and squeezes of lemon. She knew how to toast her own pralines.

Above all, just as Golden Syrup was, is and always shall be the greatest taste of my life, she adored proper maple syrup at any excuse. She associated her childhood, before she came back to England when she was six, with the sweet crumbling of maple candy in her mouth. This uber-fudge is terrifyingly expensive but given the choice between perfume or maple candy from an American duty-free she will miraculously, and to my considerable gain, choose the latter.

The only incompatibility between us on the sweet front is that I remain the same sterile-conditions man I have been since the ants broke into the syrup when I was very young. She, on the other hand, has the great British clear-your-plate tradition running through her veins too. This came to a head on our wedding day. There was an almighty panic from the caterer because the ghastly Eggwina Currie had just opened her huge gob on the subject of salmonella in eggs. Our cook was terrified that if she made the lemon syllabub my wife and her mother had devised with buckets of raw egg they'd all be sued.

Nonsense, said my wife. We'll all have the syllabub. Which everyone did. Though if you were to have looked at the top table where a man was sitting next to his new wife you might just have seen that his syllabub was taken away untouched.

Paradise Cake

My wife was pregnant. On the day my son was born I went about erecting a water-birth pool, clicking panel A to clip Z and stretching the sterilised liner across the oval base. I felt confidently ahead of the game, my wife still in early labour, as I began to fill the pool from the tap in the delivery room. Although the wise men of the hospital authority had insisted that the floor be surveyed to make sure it could carry the weight of a small pool of water nobody had thought to provide both a hot and a cold tap. The only one in the sink sprayed water so scalding that you could cook a lobster in it.

Less calm now, it fell to me to borrow a bucket and while the pool was filling up with the steaming stuff, walk up and down the corridor to the next room to fetch pail after pail of cold water so that my wife would not be boiled alive. For an hour I traipsed with my bucket, a human blender tap, until the pool was filled with lovely warm water ready for the arrival of my wife and child. I was exhausted.

All that kept me awake and cheering my wife on at the head end after that was to devour the entire bag of emergency confectionery we had brought in to sustain her, as advised by the leaflet from the National Childbirth Trust. It must have looked like a scene from the *Simpsons*, wife labouring, nurses dabbing, and husband surrounded by sweet wrappers tucking into the final bar of Kendal Mint Cake. But the truth is that sweets got me through the birth of my firstborn son and that's as good a product endorsement as Mars should ever need.

What is less easy to defend is my conduct in the labour ward afterwards. My workplace had rather brilliantly couriered a basket of fresh muffins from a trendy shop in Soho instead of a bunch of flowers. My wife was delighted and tucked in without reserve. There was milk to be made and she had every excuse.

However, the most wonderful thing about new babies which nobody ever tells you is that they sleep a hell of a lot in their first couple of weeks, a phoney war before the next five years of going bald with fatigue. So, I soon found myself alone in a side room in Greenwich Hospital with a sleeping wife, a sleeping baby and a basket of muffins.

There are only so many times you can read the nurse's *Daily Mirror*, so I made the muffins feel wanted, and when my wife looked in the near empty basket the next day I told her that I had of course handed them around to the wonderful nursing staff. She was proud of her husband's prized courtesy in his first few hours as a responsible new dad.

We drove home across Blackheath the next day at ten miles per hour. As the first of our friends to have had children and with grandparents away, we made as good a fist as we could of following the instruction manual and keeping our baby alive. Mothers fill with strange new hormones which give them the strength of ten, but fathers do too. I felt on top of the world. It was September, an Indian summer, always my favourite and most propitious time of year. I could not see how this mood of joy which seemed to exist in a tangible hide of parental strength could ever be spoiled.

I had not accounted for the jobsworth from the Lewisham Childcare Unit. We had no idea she was coming a week or so later, and when she hammered on our door we didn't really want to let her in. She wanted to talk vaccinations and to check on the health of my wife and her intrusion seemed to be obligatory.

Having taken us through a leaflet which must have been aimed at complete morons, she weighed our baby, weighed my wife and noted it down in her little red book. Then, as I sat adoringly on the double bed with these two incredible people I was to care for as long as they needed, she addressed me.

'And how much do you weigh?'

'Me?'

'Yes. You.'

I said I didn't know. Maybe thirteen stone, perhaps a bit under.

She looked me up and down.

'And what are you going to do about it?'

There was an answer I would have liked to have given about people in glass houses not throwing stones, but I didn't say anything because she had more.

'You'll find we run a well-man clinic in Lewisham now and there are many leaflets and special sessions arranged for people such as yourself.'

I thanked her and agreed to look into it the next day, which I did, to find that the well-man clinic in Lewisham ran on wishful thinking. But she left a deep mark. I was not even particularly overweight. I might have been scoring fewer goals but I was still tooling around a soccer pitch like greased lightning. I wasn't breathless, easily tired or in any way infirm. As far as I could tell I was in my absolute prime. But the unwanted and unwarranted insensitivity of that health visitor began in me a vein of suspicion of all those who attempt to dictate policy over the issue of weight.

I had known for years that the Body Mass Index was nonsense. I was five foot ten with a pretty strong build yet my target weight was eleven stone two according to this arbitrary measure. I hadn't weighed that since I was a teenager, except when I'd been seriously ill. A few weeks ago I stood with a doctor friend watching the star thirteen-year-old player at the local tennis club. I was astonished when the doctor said that the lad was a good two stone overweight.

I challenged him. I knew the boy's grandmother who had an identical build and was as fit as a flea at seventy, herself still returning volleys with ease. Surely, I said, that was just a bit of puppy fat. He tutted.

Apparently now there is no such thing. We are in the first wave of an obesity epidemic. While this might be true in some small pockets of British and American life, aren't the rest of us who are carrying a few extra pounds at risk of being convinced that we should despise the bodies we live in?

After the health visitor left that early nineties afternoon I went in to work. A colleague who would have described himself as a chirpy cockney saw me snacking to get through a night shift and said, 'The Cakeman's at it again.'

With a slightly curled lip I asked him if he had a problem with that. He was crestfallen.

'No,' he said, 'that's what some of us call you. It's really nice. You like a bit of cake and you don't mind who knows it.'

I was confused.

'Look,' he said, 'people take all sorts to get themselves through the night here. This world would be a better place if everyone else could get by on the odd slice of Paradise Cake.'

He sniffed hard, looked deeply tired, then turned away to fetch something from the tape library in the basement.

The problem with food is that one minute it's fun and the next it's deadly serious. In the wake of the presumptuous health visitor I felt indignant. But that was before Vince. A couple of mornings a week our son went to a childminder. She was Pat and her husband was Vince, an amiable man who had devised a way of life where he could sit for twelve hours a day watching various generations of *Star Trek* on satellite television. He was amply settled in such a huge leather chair he appeared to be commanding the bridge of the Starship *Enterprise*.

The way it worked with Pat and Vince was that he was a bit overweight and didn't move, while his wife never stopped moving and had barely a spare ounce of flesh on her. Some mornings I might drop our son at nine and pick him up at noon and there was no evidence that Vince left his chair between.

Pat decided one day to go on the Slimfast plan and to that end she bulk bought five boxes of the milkshake drinks you're meant to have instead of a meal. However, after the first one or two she realised she absolutely hated them.

So Vince came to the rescue. Instead of just eating pie and chips and a mug of tea from his position looking out into the universe he had pie, chips and a Slimfast drink. He did this twice a day. His calorific intake must have been huge.

For a while that made us laugh until the day Vince died instantly from a massive coronary in the week of his fortieth birthday. He was too young to face the final frontier but what he ate and what he didn't do had killed him. I was no Vince, but it occurred to me that, with him gone, I moved one more rung up the ladder.

Galway

There were two profound secrets at the heart of my life which I was able to be open about for the first time. I no longer felt alone.

The first arose from the way the world was changing to feed the needs of a man like me. I would never deny my beloved M&S custard tarts, but sometimes I walked past them like an old lover in the street. Now I had a home and a wife and an increasing quantity of children I also owned a microwave. This opened an entire new frontier. I could take the M&S Syrup sponge, their spotted dick, and a pint of custard, march them briskly up the hill from the Lewisham branch and they would be hot and ready to pleasure me less than four minutes after entering the front door.

The dilemma was that the Judeo-Christian mind comes to a point where it turns pleasure to guilt even in the non-orthodox believer. The microwaved pudding-on-tap phenomenon seemed too easy to be good for me. So did the proliferation of the new generation of Tesco Metro Stores which brought their own, inferior but acceptable, instantly heated puddings within one's daily reach.

But the great thing about my self-conscious generation was that, although we did it to a fault, we did not seem to have the slightest problem in talking about ourselves. Men were bandying tedious tales of addiction to porn but women's confessions were more interesting to me. They were defining the idea of the chocoholic, owning up to handbags full of chocolate limes and Flakes.

Every time I turned on *Woman's Hour* as I sprinkled talcum powder on to another backside there were women who were speaking for people like me. They were sharing their private confectionery secrets and I have never felt so like I was in the wrong sexual skin in my life. The relief was extraordinary.

I led a double life in the realm of food. For work I would have to sit and watch the most preposterous, pretentious, self-absorbed half-wits eat jellied coxcomb at £30 a plate in some fancy new restaurant as we met in the hope of parleying a stupid little independent production company into a major bidder for an ITV franchise. Watching them nibble on this gristly meal, which looked precisely like the Kellogg's logo, took me to a new level of despair over the Emperor's New Clothes in British cultural life. Once I'd left work, I bade my colleagues goodbye outside the Groucho Club where they pissed their young lives against the wall and went home to the domestic cliff face. I was wholly committed to family life, but every man or woman must have some means of temporary escape.

I did not have a garden shed, so instead I drove our car to the Tesco Garage on Lewisham Hill on the genuine pretext that we needed six more pints of milk. We always needed six more pints of milk. There, I would talk to Eddy, the genial Irish manager, about the way of the world, buy an *Evening Standard*, a box of two Belgian buns and a bottle of chocolate milkshake.

I would drive from the forecourt and park in a quiet side road near Hilly Fields, switch on the enjoyable babble of Radio 5, and indulge myself. It was only ever about twenty minutes, but this small freedom had the effect on me of an hour's power nap. None of it did me any good as either brain or actual food, but it restored me entirely.

This ritual stayed my secret for about a month, until our friends who lived in Hilly Fields drove past me one day, their children pointing at my mouth bursting with bun and the paper spread out on the steering wheel. They'd seen me for what I was. A few Christmases later their mother made me my own papier mâché box for holding a Battenberg cake. I felt profoundly understood.

There was a second secret which became more widely spoken about in the nineties, adoption. I had always known that I was adopted, and when our second son was born and everyone said he looked like me I thought for the first time of the importance of family likenesses.

There followed a magical sequence of discoveries and events which led to me meeting my natural parents for the first time. This should have been extraordinary enough, especially as they had gone on to have four more children, my full-blood brothers and sister. And they were Irish, which was thought to be a lucky call.

It is a shock, of course, to discover several people you didn't know existed who all look a bit like you. This undeniable similarity was rammed home further when I discovered that five of them were also as extreme in indulging confectionery pleasure as I was. My new number one brother was a very hard-working electrician who powered himself with as many as six Mars bars per day. This I had to respect, especially as he somehow did it while remaining slim and keeping his own teeth. My sister had spent her childhood with all her energy needs provided by chocolate and, like me, she had been an accomplished concealer of the evidence down the back of sofas and under the bed.

The other two brothers went weak if they could not do some hardcore confectionery every day of their lives. Most impressively, and most encouragingly for me, was that my natural father was as greedy as me and the healthiest-looking seventy-five-year-old I have ever seen.

By rights he should have been part of medical history decades ago. His favourite tipple, for example, is a neat slug of Lea & Perrins Worcester Sauce. His ideal evening meal is a large tin, and I'm talking the one everyone else only gets at Christmas, of Quality Street.

But most exciting of all, and profoundly affecting, is that when he was an apprentice in County Galway his first ever job, the start of his life as a man, was in a sugar beet factory in Tuam. When we first met he told me a million significant things about his Ireland, everything he knew, but this one story seemed to plug me into the very earth. Soon after I'd met them for the first time I'd driven through Galway following a lorry spilling unrefined beets all over the road, where they'd been run over and looked like exploding hearts. My own father knew the process of refining them into sugar, and not only that but he could build the machines to do it too.

Sadly, that beet factory in Tuam recently closed, and many others on the British mainland too. Once they gave us freedom from slave sugar. Now they are undercut by cheap beet from Poland. Another little sliver of our food security has gone, and all the technology people like my father developed and maintained will soon have disappeared entirely from our islands. As Franz Karl Achard slaved to prove centuries before, skill such as this is hard won and once it's gone may never return. Today, other than on the flatlands of East Anglia, we no longer grow the stuff so much of us adore. Soon the world will have a difficult choice. Do we refine the beet into sugar, or into bio-ethanol to power our cars? Our own national choices will be irrelevant, because soon it will all be grown in former communist Europe.

It was while my father and I were having a long and involved discussion on the subtle differences between the nougat in a Milky Way and a Mars bar that he revealed that one of his first jobs as a young Irish man in England was fitting machinery in the Mars factory in Slough.

I was flabbergasted. Did he understand the significance of this? Mars were no Willy Wonkas. They were as secretive as Coca

Cola with their recipes, and unlike Cadbury's they would sooner go out of business than allow factory tours. The paranoia probably arose from their founder having himself been an industrial spy at the Toblerone factory.

Yet my own father, my flesh and blood, had been in there, and remembers being told he could eat as much as he liked and, after the first week like everyone else never wanting to eat another Mars as long as he lived. He still won't eat them today. I was a lucky man. One father had gone to war and the other had gone to Mars.

Kulfi

I have never been a Plastic Paddy but a still youngish man may be forgiven some small pride in suddenly being a Celt, a Joycean, a cracker of jokes. Yet there are clear disadvantages too in drawing from the Irish gene pool. Every now and then it throws up a six-foot-something Liam Neeson with a huge strong conk and the purr of a lapsed priest. Most, though, seem to have rectangular frames with a blunt prettiness and soft sometimes pudgy out-lines. Wayne Rooney is, after all, one of us. Improvising an emergency piece to camera in the absence of a presenter at the Cannes Film Festival one year, I realised that, in the harsh sunlight of the south of France I now looked like Eamonn Holmes.

I thought I was at ease with my caked-up life and my swelling midriff but I was not fully reconciled to how this made me look in 625 lines of horizontal analogue video. When I saw the rushes in the smelly edit suite built in a Winnebago with a hole blown out of the side I ordered the editor to hand them over. Later that night I threw them into a bin outside the Carlton Bar.

The next telly job changed my perspective. I was the producer for some programmes from the Indian Film Festival in Banga-lore. The Indian cineastes are so keen and their state broadcaster

Doordashan so broke that they interviewed anyone with a vague connection to the festival and then broadcast what we said without a single edit until the tape ran out.

With memories of my sweating appearance in my Cannes monkey suit still fresh I was reluctant to appear, even though everyone ought to look good dressed in linen in the mimosa light of an Indian afternoon. But there was no polite way to refuse, so I forgot the millions watching and delivered my old tat about the Indian Festival from the British point of view, sharing with Mother India my wisdom gleaned over the three days we'd been there.

I saw my interview transmitted on Doordashan TV in a beautiful cake and ice-cream restaurant in Bangalore. We had been shooting a location report with the great Keralan actor Mamooti and he had kindly taken us all for a kulfi, an almond ice cream that was thick and gritty and wonderfully sweet. The only aspect wrong with an afternoon like a paragraph from Salman Rushdie was that our presenter couldn't eat anything, as Mamooti had punched him in the mouth by mistake when we shot a scene showing how to stage stunt fights.

Indian actors understand image, vanity and the potency of fame better than any civilisation in history. They are gods, and it amused Mamooti that I was so uncomfortable with my small moment in the sun. I explained to him that it wasn't just the unedited nature of what I was saying, it was the way I looked these days.

He and his entourage simply couldn't understand my point, and as I thought for a moment about what passes for a good-looking male lead in Indian cinema it dawned on me that my neurotic moaning about my boneless moonface could have been applied to every one of them.

Mamooti was a producer too, and a director. Many of his kind become politicians, some ruling over entire Indian states. He was too nice a man to move out of film, but in that world he was an absolute king-maker. As we drank tea after the kulfi he explained to my amazement that there was a part he had in mind for me later in the shoot, playing a handsome Westerner. I was completely taken aback and asked why on earth he would want me. He moved the palms of his hands down his face and around his torso as if he was describing the curves of Miss India.

'Because,' he said, 'in India you look like a healthy, wealthy and very successful man.'

Back in London I entered our Soho production office about half a stone lighter than when I'd left. I hadn't eaten any meat all the time I was in India, and apart from the calculated risk of the kulfi, I hadn't had any animal fats at all. I had a bit of a tan too and, walking through the desks, I felt as if I was being wolf-whistled by the staff. You look fabulous, they said.

Fabulous in India for being fat, and fabulous in London for being thinner. I couldn't make any sense of it.

Diet

However much a man is reconciled to the reality that he is never going to resemble a consumptive poet, every now and then he is overcome by the urge to try. The most common spur for me was the possibility that I was going to see old friends who hadn't seen me for a decade at a reunion. For a long time I trimmed down just enough to get away with it so long as I wore a baggy jacket, but there was one meeting with a dear old friend whose parting words to me were, 'you can't just keep buying bigger clothes, you know'.

The difficulty was that as a sceptic in the field of weight-loss every single idea always seemed to border on the insane. Gay men at work were the best source for this type of thing, and it was a from a young man called Dougal that I first heard news of the grapefruit diet.

'It's brilliant,' he said. 'You've got to have a great big plate of egg and bacon for breakfast and you can eat as much as you like so long as you drink half a pint of grapefruit juice with it.'

Dougal explained that as far as he understood it the science went like this. The egg and bacon were protein and the grapefruit juice was an acid so, as long as you got it all down the hatch at the same time, the grapefruit ate your breakfast before you did.

His huge gang of friends swore by this as the way to shift an emergency stone before taking to the streets in their thongs and nipple clamps for the annual Gay Pride march.

I tried it for a week, lost a few pounds and grew a colony of zits the like of which I had not seen in the mirror for twenty years. Worse, like Dougal and all the friends he introduced me to who swore by the scheme, I exuded a smell like rotting meat from every pore.

After this came the no carb diet a big fat solicitor showed me. We had a meeting one lunchtime and he took an entire silver salver of sandwiches which might normally feed a room. With surprising delicacy he then threw away the little triangles of bread and swallowed all the meaty fillings. The last time I saw him he was struggling with the stairs.

Eventually an image presented itself to the public imagination which felt to me as if it required my investigation. Nigel Lawson, formerly Chancellor of the Exchequer and one of a number of ex-ministers who'd had hissy fits about Mrs Thatcher, disappeared from public life and came back a year later looking like a deflated Zeppelin. This was just before we came to know of his daughter, the divine Nigella, who was to become a figure of inspiration to those of us who valued cake in our life. Then she was famous only for having the most ridiculous first name since Monty Python called a father and daughter Brian and Brianette.

No matter what one might think about her father and his views of the European exchange regulatory mechanism one had to take one's hat off to him in the field of weight reduction. He had once resembled a living Toby jug, and now looked as if he'd had all-over total liposuction. With a leftover book token I bought a copy of his diet book, which, like the new him, was slim and blue.

The opening pages were a feast of alienation for a man with four children and a chaotic life. First of all, it was clear that the lazy oaf could hardly boil a kettle and that the excellent Mrs Lawson did all his cooking for him. Second, when she wrote in her introduction that she was effectively his galley slave for the entire period of his shrinkage and moreover that 'Nigel's absolutely favourite food is grouse', I realised he and I could never sing from the same diet sheet. There wasn't a lot of grouse in Lewisham. Goat, yam and red snapper, but no grouse.

Yet despite my encounters with such absurd weight-loss plans and my not being *that* bothered about how much I weighed in any case, I kept having a go, always as pleased as a young bride if the scales showed four or five pounds had disappeared. After years of less than diligent experimentation it seems the only programme that gets results is eat less and move more but who on earth ever does that?

Hemp

The main benefit in clocking up the years is that eventually the world catches up with you. Unfortunately, just as you appear to have achieved some harmony with it in one aspect of life, another starts to pull away in the opposite direction.

So, for the cake-lover at large in the new millennium there was no longer a need to feel like a freak. America was in the vanguard of cake liberation. In New York, cake culture was revived so powerfully that there was a huge lawsuit between rival shops (called the Cup Cake wars) over who could claim the credit for this. The Magnolia Bakery shop in Bleecker Street and the Little Cupcake Shop in Brooklyn slugged it out through the courts, and in the process spawned an upsurge of rivals across America.

As if lawyers heating the ovens was not enough, Carrie Bradshaw and chums then came over all cup cake in *Sex and the City*, leading to huge queues of fans after identical pâtisserie outside the Magnolia, who were now benefiting from both the greedy and the pink pound. I wasn't pink but was undeniably greedy and, away from the media pundits labelling this a TV-led phenomenon, ordinary New Yorkers reclaimed an honest pleasure from their childhoods and shared it with their own grateful children.

Further upscale on both sides of the Atlantic there was an outbreak of chocolatiers providing the rarest cocoa products from named plantations in specific south American mountain ranges. The suburban git in me despised this and all the posturing that accompanied it, but when I tasted the cocoa solids from somewhere in Venezuela compared to somewhere else in Chile I felt like a budding wine-buff. I would never betray my own preference for bottom-of-the-heap Kipling through Cadbury mass-market junk, but even someone as unreconstructed as me could see the genuine pleasure of those who chose to go to the top of the range in sweetie world, and the pioneers in this field won my sincere respect.

I was glad to see what had once been my private convictions enter the culinary mainstream but as it became newly respectable to be a confectionery enthusiast, elsewhere unpleasant individuals were working to reinforce taboos over carrying the extra pounds this inevitably led to. With my old friend from the Blackheath basement I wrote a comedy pilot called *Safari Supper* about a terrible moving dinner party which trails from the first course at one house to the ghastly dénouement over port and cheese somewhere else.

We delivered the script just a couple of years ago, and were very keen to have one character so cold-blooded that it was plausible, as the evening wore on, for her to ditch her spineless husband and begin an affair with a female neighbour. By the time we'd finished inventing her she was to be a morning television doctor with a greater appetite for her next close up than Gloria Swanson. To dramatise her incredible cruelty, egomania and insensitivity we decided she'd be working with an overweight family who we amusingly called the Atkins who sweated away in the gym on a fictional *Sunrise TV* every day while she flogged them on with her cruel words. We called her Dr Carol 'The Body' McPherson.

To leave the viewer in no doubt about her incredible ruthlessness and lack of empathy for others, we came up with what we thought must surely be the medically unethical idea of Dr Carol placing the Atkins behind tables groaning with the food they ate each day. We then had her make them weep openly, especially the younger members of the family, by removing any foods she considered unsuitable and railing at them about late-onset diabetes and childhood obesity. Eventually there would be a falling out with the Atkins dad when Dr Carol told him that if his porky son wanted a sweet treat he should have a paw-paw or a pomegranate and Mr Atkins yelled at her, 'Right, they have those in the vending machine at school.'

The whole point of this horrible doctor was that she had crossed two lines which thousands of proper doctors would never traverse. She would do anything for fame, including bending the medical message to the fad of the day, and she would abandon her duty of care to score cruel points off vulnerable patients. We felt she worked quite well and she certainly made us laugh.

The pilot script entered development hell where it probably languishes even now, but the really hellish aspect of this was turning on the television a year later and stumbling across a programme called *You Are What You Eat*. It was presented by a woman calling herself Dr Gillian McKeith, although she was not medically qualified. It was as if the Comedy department at Channel Four had put our script in a dustbin only for it to be retrieved by the Factual department and made into a series of documentaries.

As far as I could see this programme was the licensed bullying of fat people who were made to stand behind tables groaning with pizza and other refined products they habitually enjoyed. Its extraordinary unique selling point was that in order for McKeith to prove her point she sifted through their excrement. This had been way beyond our comic imaginations.

Taking soundings at dinner parties is not empirical research, but when I raised the subject of what seemed to me a freak show the table usually divided between those who felt we had a new witchfinder general in our midst and those who had begun to eat the McKeith way.

Almost as a reaction to this I went for the first time in a decade on a trip down memory lane to the Harrods Food Hall, pointing out to my wife where I'd started up the Jelly Belly counter and the cabinet from which I'd served my first cream slice. We were approached by a young man who would have been me twenty years before and invited to try something that looked like a little piece of flapjack on a cocktail stick.

I bit into it and was inclined to deal with it as the journalist Jonathan Meades had once dealt with a Linda McCartney sausage, to spit it out into my hand. But I didn't have his upper crust balls so I swallowed this sorry item and demanded of the young man what on earth he'd just persuaded me to eat.

He handed me an unopened Dr Gillian McKeith Hemp Bar. I could not believe this was masquerading as food. You might make a pair of wholemeal shoes with it or mix it with wattle and daub to recreate a medieval house, but nobody would ever have a genuine desire to eat it. Wasn't the whole point of food that it tasted good on the way down? Was this the future for those perceived as overweight, a prescription for a hemp bar and a daily admonishment from the likes of McKeith?

I turned the bar over and there on the back was a small photograph of its originator, like a horrible 3-D image from Tom Riddler's diary in *Harry Potter*. I looked around the Food Hall with all its wonderful produce hanging from the ceilings, presented as beautifully as anywhere else in the world, and wondered if perhaps my patent peanut and milk diet had been such a terrible idea after all.

Ballast

There is more etiquette in adult male behaviour than meets the eye. A man of my age must be careful not to refer to 'our generation' with a man who has only just turned thirty-nine. He might look exactly the same as me, but it is crucial to his identity that, while he was enjoying Duran Duran doing his A Levels, I was hating them in the final year of my degree.

This need for care extends to weight relativity. I have been on both sides of the fence, having given offence in comparing myself to a fellow diner in denial or being deeply offended when an utter fatso talks to me as if he and I were the same. I always want to say I might be a walrus but they are a whale, but I never do.

There are ups and downs. A minor public-school toff working in a tailor's in Exeter told me through his brown teeth that the only neck he had ever measured bigger than mine belonged to a prop forward for Bath Rugby Club. I nearly took his tape measure and threaded him on it.

On the other hand, there are many people as I get older who hold with Caesar's dictum:

> Let me have men about me that are fat
> Sleek-headed men such as sleep o'nights

> Yon'd Cassius has a lean and hungry look
> He thinks too much: such men are dangerous.

So that, rather than entering a meeting wanting to apologise for one's plumpness, it is possible now to go in looking literally well rounded, a safe pair of chubby hands. Amazingly, this seems to hold with the opposite sex's perception. If they can't pull the tennis professional, what could be safer than an XL man whose body is no longer under guarantee, though I should make it clear that I have no hands-on experience of this.

Some sports welcome the big hitter while others do not. I still rollock around a football pitch once a week, and like every other man of middle years I use my sprints sparingly. If I do make clean contact with the ball it flies into the net like a ball of fire, impelled by an almighty blow from fifteen stone of red meat.

I have played for a team called the Cakemaker's Dozen for twenty years now. I didn't choose the name, an obscure tribute to my boyhood hero, Derek Hales, who shared his name with a now forgotten cake brand which once rivalled the mighty Kipling. Playing for them recently on a Devon pitch I was tackled by a great lump who would formerly have crushed me like a butterfly. He flew off my brutish shoulder and somersaulted into a ditch. This weight advantage can apply equally in both tennis and golf, where there is no substitute for a well-timed stroke backed up with brute force.

Two sports came into my expansive life too late. For some years now we have driven our rusty people mover (200,000 miles on the clock) into Europe for an ultra-cheap skiing holiday. My wife and children zoom off down the slopes like mountain goats as soon as they've hired the gear. On my first time skiing, aged thirty-nine, I was as ever sixth in the pecking order in the ski-hire shop and was issued with a pair of size 7 boots for size 9 feet.

On my first morning I felt as if I was developing gangrene, and after a couple of disastrous lessons with a young French bastard called Thierry I realised I had left it too late to start from scratch.

I had an irrefutable conviction that if I gave myself to the slopes it would prove conclusively that the bigger they are the harder they fall.

I resigned myself to service as chauffeur and chalet maid and, though I lament each time we are in the Alps that they are covered in snow rather than edelweiss, I enjoy the pleasure of others and watch the rescue helicopter from the safety of a bar.

Socially I am in my element. I always take the lift to some distant cabin in the Trois Vallées my party has skied to and share the same huge pasta and pizza lunch, washed down by half pints of hot chocolate and piles of waffles. Everyone else returns from skiing leaner and with an orange tan but I return through the Channel Tunnel fatter and paler. Unlike a huge percentage of males my age, though, I retain intact knee ligaments.

Sailing came too late as well. My children could probably find their way across the Atlantic by now, and passed the Royal Yachting Association First Grade test without blinking. I found that the titchy yacht I was asked to steer around the sea off Marmaris in order for me to pass the same grade simply wouldn't move in anything less than storm-force wind, when it immediately capsized. Its hull was so small and I was so big within it that I found it almost impossible to turn round, let alone dance like a flea tacking and gibing and generally avoiding the inevitable blow to the head from the boom.

Only once did I feel at ease on a yacht, and that was a big one being raced in a regatta by an extraordinarily competitive friend. I explained to him that I didn't know port from starboard and the only knot I could secure was my shoelace. He didn't mind, he said, he had a role just for me.

We were half a mile out into the ocean when he explained what this was. He wanted to use me as ballast. After a few minutes crashing through the waves he stopped calling me Paul and called me Ballast instead. My purpose was to fling my weight from one side of his blasted boat to the other as he twisted it round buoys screaming, 'Give way!' to other mad sailors, with a committee boat parping its outraged horn behind us.

Afterwards he bought me a drink and said with complete sincerity that I was amongst the finest Ballast he had ever worked with. If I'd been a girl I think he might have kissed me.

Elsewhere in daily life a lot of things close in when you've become a shape-shifter. Most people remember returning to a childhood haunt that once seemed like a jungle and turns out to be a tunnel under a rhododendron bush in the park. If your waist expands by a foot in under a decade airplane seats shrink that way too. On double-decker buses your shoulders seem to brush the sides going up the stairs. When you go to theme parks with your kids and a safety bar is slammed across your lap you find it only just clicks shut. If you try to squeeze behind another chunky person in a village shop there is no longer clear passage for two. Yet all this can give the feeling not that you're bursting out of your skin but that you've grown into it.

Having more than a medium dad is fun for children. Dive bombing them in swimming pools can start a tidal wave. When you carry them around you have a handy shelf. If I say I'm thinking of losing some weight my daughters immediately cuddle up as if I am some kind of huge animated teddy bear which, since they don't remember any of my former glories, is probably exactly what I seem.

Keeping in touch with the new products on the cake and sweet shelves yields a complicity with children. Recently two classics from the genres of cake and sweet were combined in a package so garish as to be irresistible. The Refresher Cup Cake was launched, its box shining in the vividly striped pinks, blues and greens of a tube of Refresher sweets. It deserved a place in Tate Modern. The cup cakes themselves were topped with either pink or yellow fondant icing and the brilliance of this topping was that it fizzed precisely like a Refresher.

I scored a couple of boxes of these privately but was then so enthused that I shared my secret with the family. Reverently they came with me to the Spar shop and over tea they unanimously agreed that they were absolutely repulsive. My wife was delighted, assured at last that the curse of Kipling-worship has stopped with me.

I was so excited about my find that I told friends about it too, and I believe some of them may even have followed me into the faith. What worries me terribly is that some of my Refresher converts will have devoured the box and then hated themselves for it. Of course, for those who are genuinely bulimic this would be awful. But I am concerned that the rest of us greedy guts who love eating sweet stuff now think of our honest greed with self-loathing, when all we are doing is scoffing a packet of cakes. Or worse, that our obesity epidemic-obsessed culture is forcing us to think in such terms.

I would rather we see the Refresher Cup Cake for what it is. It is an end product of an industrial process which began in the nineteenth century with the great philanthropic sweet manufacturers and now – with the merger of great names Cadbury's, Trebor and Bassett – allows a diabolically brilliant synthesis of the very concept of what is a cake and what is a sweet. It is the modern world and we must deal with it, but there is no point blaming the corporations or our parents or our brain chemistry if we are seduced by them. We just are.

So many therapies and religions follow the line today that we must love the sinner but hate the sin. For the average Bunter who has simply forgotten to say no this is a moral and mental catastrophe, a symptom of the new puritanism. Eating cake isn't a sin, neither is scoffing so much cake that you put on a few stone. It is an indulgence. You cannot hate an indulgence.

My childhood was filled with role models who proved that you could be fat and functional at the same time. Bunter was the only example of the bad egg from the breed, and that wasn't because he was fat. It was because he was a cad, a sneak and an all-round slippery eel. But nobody could say that of Friar Tuck. With one hand he could marry Robin Hood to Maid Marian or wield half a chicken, and with the other smite the Sheriff's goons with a broadsword. He was fat, fit and wise. Similarly, the TV detective Frank Cannon, played by William Conrad, had to put up with being called Fat Man by West Coast wasters but he soon had them cuffed.

On prim little programmes like *Blue Peter* and *Newsround* today, however, furiously ambitious presenters with pierced navels ram home the latest food paranoia from the Ministry for Spoiling Childhood. With consummate hypocrisy they report that inactive children are watching too much TV. Then they frown about processed food, so that in addition to terror of child abusers, world destruction and climate change their viewers have to worry about sweetie day too. They do not understand that nobody ever gave up anything someone told them was bad. And if it's not bad, if in well-judged portions it is actually good so long as teeth are brushed before bedtime, self-proclaimed authority figures must be aware that by crying wolf over sugar a generation may not heed them at all when there really is something grave ahead.

Hershey

If I could ask for a facsimile copy of any newspaper in history I would choose the *New York Times* for 9 November 1923. On one side of the front page is the dreadful news that four years after one war had ended the cause of the next one had begun his inexorable rise. Adolf Hitler had staged his putsch in the Munich beer halls and was preparing to march on Berlin. The most tragic part of the article is that it features many quotes from concerned Germans begging someone to stop him or risk the ruin of Germany.

In the next column is a story which reveals by juxtaposition the difference between the dark heart of Europe and the common decency of the United States of America. That was the day when Milton S. Hershey revealed he had signed over his entire fortune to fund a school for the benefit of poor orphans. The trust he set up is now thought to be worth in the region of $8 billion. It was the greatest act of educational philanthropy in the history of the world. Under the cathedral-sized dome of the school today this page of the paper is blown up as big as a man for all to see.

That Hershey could afford to make such a noble gesture was due to the success of his eponymous chocolate bar, and later of his Hershey Kisses and Reese Peanut Cups. Hershey is an

inspiration to any romantic. Despite being raised in the Menno-
nite community of Pennsylvania he was not an orthodox believer. As such he was an entirely modern figure for the twentieth
century.

He went bust many times trying to get his sweet-making
business off the ground and was only saved by a wonderful
intervention from England. His only small success was making
caramels (or toffees as they are known in Britain) and his idea
had been to make them less sticky and more solid by increasing
the amount of milk they contained. This afforded the Hershey
Denver-recipe caramels an unlikely cachet as something approaching a health product, and a British importer passing
through Lancaster County discovered them and thought them
good. He placed an order on the day Hershey was about to be
foreclosed at the bank, with the promise that if they arrived in
London in good condition there would be more orders to come.
They did and Hershey expanded from that day forward, his
struggle relieved by a stranger passing through town.

Curious, he hired an agent to find out what the British were
doing with these tons of his caramels. He discovered that they
were covering them in milk chocolate and selling them as
chocolate toffees. He knew nothing about chocolate but set
about teaching himself the process in a lab he set up at home.
The recipe he settled on came entirely from trial and error, and
it's thought the reason the British loathe the Hershey bar so much
is because of its inventor's damaged taste buds, half-kippered by
pipes and cigars. In Britain we are used to the smooth slow
release of cocoa aroma and taste in a melting texture from
Cadbury's. Hershey bars give us an unfortunate first impression
of rancid butter followed through by a bitter and acrid aftertaste.

Having made himself rich, Hershey then shared his good
fortune with most of Lancaster County, and built a model town
in which to house the thousands of workers on his production
line. To see this town now, spread out across Amish country in a
green swathe of Pennsylvania, is to perceive the difference

between American philanthropy – expansive, seed-sowing, unconditional – and the more proscribed communities at places like Port Sunlight near Liverpool. There, the soap magnet William Lever ruled every aspect of a worker's life from what he wasn't allowed to drink to what he couldn't do with the opposite sex.

In response to the terrible hardships of the Great Depression Hershey refused to shut his factory but instead expanded and hired extra hands to build a theatre, a hotel and a sports stadium for the town. He truly believed that the profits of his good fortune did not belong to him and that it was his moral (not religious) duty to prime the pump by reinvesting.

When news got out about his school his chocolate bars had the unlooked-for advantage that by buying one you might be doing some good, and throughout the Depression sales stayed steady. This wasn't just because the buyers were backing a good cause. In thousands of cases these small blocks of sugar and cocoa saved city dwellers with scarcely a cent to their name from starvation. Just as in post-industrial Britain, an essential carbohydrate.

In the past few years the Hershey Trust witnessed the most terrible boardroom battle, with one faction keen to sell the company to Nestlé and the other desperate to keep the company independent to protect the town, the factory and the school. What played out was like an updated take on *Wall Street*, and for once the good guys won, despite the blandishments of the ranks of accountants and management consultants after an all-time record fee.

Milton Hershey is my hero, and his personal life followed an improbably cinematic course. He adored his wife, but they didn't have their own children and she died very young with what we would now recognise as multiple sclerosis. This was why he endowed the school. In Cuba he was concerned both to secure his supply of cocoa and sugar and to save his workers there from exploitation, so he built whole towns around his plantations and ran them on the idealistic lines of Hershey, Pennsylvania. And in a twist which is so bizarre that it beggars belief, he and his wife had tickets to travel on the *Titanic* but, owing to a business complication, they sailed home three days earlier. Without that non-divine intervention there would not be a town or a school or possibly even a chocolate bar today. I admire him because his creed was post-Christian with all the nonsense taken out and all the good intentions put into action. And as I get older each time I go to America I find that, with the ageing of my own taste buds, I even like his chocolate too.

The last time I went to Hershey was in what is now referred to as the post-9/11 era. I was driving across the border at Niagara Falls from Canada with the intention of visiting the town and its local history museum. At the checkpoint into the USA, as a single man in a hired car coming into the country by land, I was selected for interrogation by the Department of Homeland Security.

I parked the car and entered a low grey building where after a nervous wait I was interviewed by a beefy guy with a gun and a computer he seemed only just about able to switch on. He asked me the purpose of my visit. I told him I'd been doing some documentary work in Toronto and I was just dipping down into Pennsylvania for a couple of days to visit Hershey.

'Where?'

'Hershey.'

'Like the chocolate bar?'

'Yes.'

'Hershey, PA?'

'Yes.'

'Why?'

It was a reasonable question. I told him that I was fascinated by the town and planned to do a bit of exploratory research.

'You want to do research about a chocolate bar?'

I started to sweat. It did seem an unlikely reason to enter America. But he seemed minded to press on.

'If you say so. What is your destination address in Pennsylvania?'

'I don't know. I was thinking about a Holiday Inn Express or something.'

'Look, guy, help me out here. You can't give me a destination address, you don't get into my country.'

We looked at each other dumbly. I didn't have a clue what to do.

'I just need a zip code. You got a zip code for me?'

'I don't know. How would I know that?'

'A street name. You know a name of a street?'

I thought for a moment and realised that to my surprise I did.

'I know two street names in Hershey,' I replied. 'There's one called Chocolate Avenue. And another called Cocoa Street.'

'You serious?'

I assured him that I was. He punched in some details with his meaty fingers and to his obvious astonishment there they were.

'What do you know? It's giving me a zip for both. We'll enter you for Chocolate Avenue.'

Somewhere in the Pentagon this will remain on file for the rest of my life and if the war on terror hits Hershey I'll probably be extradited for questioning. But I hope nothing bad ever does happen to this Pennsylvanian paradise because in these troubled times it's the town which made me fall in love with America again.

Sachertorte

While Milton Snavely Hershey was emerging as a hero from the *New York Times* of 9 November 1923, across the page Adolf Hitler was reaching towards his own historic destiny. His actions that day presaged decades of pain, with the consequence of rendering my parents' generation embattled, bombed and rationed. Hitler's homeland took much longer to recover than mine. The first failure of his life was as an artist in Vienna after the World War One, the city of Klimt and Schiele. Within twenty years he had annexed Austria and led it to defeat within the Third Reich, leaving the Hapsburg jewel of Vienna ruined and divided between the Allied victors.

One of the most vivid illustrations of this ruin was the Hotel Sacher opposite the Opera House. It was used for many years after the World War Two as a billet for the British army and secret services. Where high society had gathered to dine before the Opera Ball the British cavalry kept its horses and occasionally improvised a game of indoor polo. Graham Greene stayed there in 1948 and wrote much of the treatment for *The Third Man* in an upstairs room overlooking the devastated city, urged on by the director Carol Reed.

The most heinous loss for Vienna was its entire population of

Jews. Freud escaped in 1938 through pressure from both the American and British governments but if he had attempted to leave after the outbreak of war he would never have made it. This was the fate of the rest of Viennese Jewry, and in a city which still struggles with issues of racism this tragedy is honourably marked by museums and works of public art.

There was one aspect of Viennese life, however, which survived. It carried along with it the best European tradition of free thought, the use of conversation and language for their own sake. It is not accurate to describe it as an establishment, because it was really an idea which happened to be shared in a hundred or more buildings across the city. The idea was *Jause* and the setting was the *Kaffeehaus*.

Much is lost in translation. Jause happens at four o'clock every day in Vienna but it is not the same as the English tea. Jause does not have the whiff of the maiden aunt; it is not about scones or cucumber sandwiches. It is about cakes and coffee or hot chocolate. It is highly unlikely that any great movements in British thought have been launched over tea, but for three hundred years Jause has been when the Viennese thought and shared ideas.

The setting is perfect. The great Viennese cafés are more prominent than any olde tea room. The Café Central, former haunt of Trotsky, has a Gothic vaulted ceiling under which it would be embarrassing to talk trivia when there is the Marxist dialectic to consider. It remains favoured by writers. The Café Landtmann in any other city would have been bought by McDonald's and made into a mega-burger outlet, exploiting its position near the broad sweep of the Ringstrasse. But the ghost of Freud is tangible, and it is now haunted by politicians from the City Hall across the road, and by actors from the Burgtheatre next door.

Taking Jause at the Demel Café is like entering a Hans Christian Andersen story. There is always a contemporary political story being satirised in large marzipan figures in its window, and in the back rooms former marzipan stars such as Kofi Annan and Bill Clinton grin at the customers sweetly. The lobby is wood-panelled and mirrored, with extraordinarily beautiful waitresses wearing a severe black uniform inspired by the convent-school girls who once served there. They seem to be playing the part as if that night they were on stage at the Opera. Everywhere there is blazing colour from the formal arrangements of Cleopatra Torte, Punschtorte, and Annatorte in glass display cases.

At the back of Demel's the kitchen can be seen through a huge glass window, and two-metre apple strudels steam from the oven, measured with a standard length of wood and cut precisely with a hot knife, sprinkled with sugar and mounted in a paper wrapping. A few moments later a hot slice goes to the table where it fuels a debate on the unconscious mind.

At Demel's you should always try the Sachertorte, as you should at every other Kaffeehäuser in Vienna, because each one makes the greatest of Vienna's cakes to a different recipe, reflecting the tortured legal battle for its name. Only at the Hotel Sacher can you buy Sacher Torte (rather than Sachertorte) and it took seven years of courtroom argument to decide this.

In the mid-nineteenth century von Metternich was the political master of Vienna, and ordered his chef to bake a great cake for a party. The chef fell ill and the only man left standing in the bakery was his sixteen-year-old apprentice, Franz Sacher. He stepped up to the plate, and composed a cake with the main and brave element of chocolate, thought then to be far too heavy and masculine for a cake in a city which usually prized fluffy lightness. Nobody knows why, but he decided to make the best chocolate sponge he could, smear it with apricot jam, and invent the chocolate icing to go over it, which gave the cake its then revolutionary chocolate sheen.

It was the making of young Sacher, who was then summoned to work for a Hungarian prince in Budapest, returning in glory to the forerunner to Demel's, then the official bakery to the emperor. Soon he set up a shop under his own name and took the Sacher recipe with him. At Demel's, however, they kept making it after he had gone.

Over the next hundred years, his son opened the eponymous hotel, made internationally famous by his cigar-chomping widow, Anna Sacher. Hers was the only place to be on the opening night of a new opera beneath Vienna's first electric chandeliers, and she was immortalised in contemporary waltzes, songs and operettas.

After World War Two the Sacher was reduced to being a military barracks and, when the Allies had gone, it reopened for business in a devastated economy. Every business had to find whatever edge it could, and so the Hotel Sacher decided to sue the Demel to prevent it using the Sacher name on its chocolate cake. The court battle gripped Vienna as much as any opera, with tales of recipes stolen from safes and intrigues involving the long-gone aristocracy of the Austro-Hungarian Empire. In the end a solution was reached. The certified Sacher Torte could only be made by the hotel's bakery while Sachertorte could be baked by anyone.

Only in Vienna could such semantics over cake matter so much, a sense of identity be defined by it. Only in Vienna could a sixteen-year-old become a national hero because he'd added apricot jam and a glazed chocolate icing to the top of a sponge. In Britain we would make a fatuous comedy from the story, but in central Europe it would be a metaphor for how a city enshrined the ownership of a key part of its culture and recovered part of its soul after the horror of war almost destroyed it.

There is, however, a problem. On my last visit to Hershey I could hardly have found a more thriving town, a factory with more secure employment, a better-heeled school. If it is not there still in a hundred years I will eat one of Milton S. Hershey's collection of native American headdresses. Unfortunately, the institution of the Kaffeehaus is not so secure.

The labour costs in making so many cakes, most by hand, are

exorbitant, four times that of a bakery back in Pennsylvania. The rent and rates for the Kaffeehaus are rising with the international profile of Vienna. Only recently it was a backwater but now it is home to establishments such as the International Atomic Energy Commission. Its geographical location at the former edge of the Christian world, within touching range of Islamic Europe, makes it a natural home for world diplomacy. These days it is rare for Vienna not to be hosting a key conference. There is talking, but it is in corporate facilities between delegates who never leave the hotel.

Each year a few Kaffeehäuser disappear, taking with them centuries of history. Meanwhile, Vienna has its first Starbucks near the Opera House. You don't have to be a snob to be dumbfounded that anyone in their right mind could go there rather than one of the traditional alternatives but eternal vigilance is necessary. If we were ever to lose Viennese café society we would have lost one of the world's most important cultural templates, as if all the cathedrals in Europe suddenly stopped saying mass.

That we might learn from the model of Vienna would be my hope for British café life. We have far too many equivalents of Café Nervosa from *Frasier* or Central Perk from *Friends*. We are not Seattle and we should stop being so skinny latte about everything. Insecure that we don't have enough sunshine to sustain the café culture we have seen in Italy, we load our pavements with tables as soon as the thermometer creeps above 15° centigrade. Ashamed of our crumpet and Battenberg past, we dish out panini and bagels instead. This means nothing to me. (Oh, Vienna.)

We need to go back to the central European model, to relearn the interaction between cake and thought, to unashamedly bring it all indoors in the largest spaces we can find. The proper café should be about huge rooms and newspapers and conversation, not sofas and i-Pods and odd little meetings away from the office.

Hershey tells us how to run a confectionery town and do great public works with the proceeds. Vienna tells us how to eat and talk in the afternoons and recover from war by the benediction of a perfectly made cake. We must hope that its survival may be assured by the example of institutions such as the remodelled Hotel Sacher, where in a spirit of slight self-parody the indulgent can take cake for breakfast, lunch and dinner and have a chocolate massage in an ultra-modern health club in between. The Kaffeehaus may need civic support to survive changes in the economy, but it would be tragic if what could not be destroyed by war was lost because of property speculation and new money flowing from the east.

Franz Sacher's patrons would have known what to do. Host one of those boring Congresses on European Enlargement and commission the bakers of Vienna to create an Enlargement Cake. European Cake Commissioner, now that would be a job for me.

Bake Off

It has taken me a move to what many metropolitans might regard as beyond the pale – the great British countryside – to remember how to live. A few miles from where I am today a famous TV chef poaches badger casserole in cider for his supper and puts it on the telly. I have graduated belatedly into rustic savouries too. The butcher's own steak and stilton pies, the local hippies' Spelt Bread, or a piece of Dorset Knob with cheddar.

The kids have learned to line-catch mackerel, to gut them on the beach and cook them over an open fire. If a neighbour gives us a brace of pheasant we don't sling it in the bin but prepare it in one of a hundred delicious age-old ways.

This will never be the real me, though. I will always be bound to sweetness. A few months ago we were meant to be at a splendid local party. It was held to celebrate a wedding anniversary, an eighteenth, a twenty-first and a couple of fiftieths all at the same time. I was forced to arrive at this magical evening, which looked like a scene from *A Midsummer Night's Dream*, late, hot and starving. The wonderful hostess assured me that a plate of food had been put to one side for me, but I simply couldn't find it. The caterers were packing up the kitchen, the hosts were dancing, and there seemed more important things for me to concern them with than a missing plate of food.

However, I saw a sight on the far side of the marquee which made me rock back on my heels. Somehow I had gone through my recent life without coming into the presence of a Chocolate Fountain. I walked up to it slowly, as if I were approaching Helen of Troy, and received instructions from a young woman in how to take a skewer, load it with strawberries, doughnuts, marshmallows and grapes, and then plunge the entire thing into this cascade of hot melted chocolate.

I gurgled with pleasure like a jowly dog slobbering on a bone. Soon my skewer was bare, and I didn't know what to do. I was still hungry and this was all that was left to eat. I'd have to have another go.

Not wanting to appear a complete pig, I explained to the attendant as I went in hard with my second skewer that I had performed heroic acts of manual labour which had made me late that evening, and not eaten anything since lunchtime. She accepted my excuse, but perhaps I should have taken care to deliver it more widely, for I became aware of rumours that I was making repeated approaches to the waitress by the chocolate fountain. I hadn't even noticed her looks so enraptured was I by her machine but she must have noticed me, the man dribbling chocolate down his dinner jacket. Yes this was lust, but it was for her fount of melted chocolate.

Chinese whispers are a satisfying part of Devon life, and it wasn't long before I was being told over dinner about this sad old man who'd been bothering waitresses at someone's chocolate evening. I was forced to declare an interest but my explanation was entirely plausible. Our guests knew my passion for the brown stuff. Indeed, they were eating my Lake Michigan chocolate fudge cake, my take on an American recipe and in my own estimation a triumph.

I was boasting about this a few days later to a friend and he said it couldn't be that good, because his wife made the best chocolate fudge cake in the world. I mentioned this vain claim to my daughter, who mentioned it to his daughter, and before I could backtrack I found that I was entered for an official Bake Off to be held and formally judged.

With respect to my opponent I was quietly confident but, always wanting to leave room for a Plan B, I left an entire Saturday morning to make the cake. It was as well, because I panicked like a novice being sworn at by Gordon Ramsay. Just as I was about to start I was advised that my electric hand-mixer had been dropped on the floor the week before and could now only be switched on and off by controlling it from the mains. I was not told that it only had two settings – off and ultra fast. Neither had anyone mentioned the smell of burning from the plug.

So before I'd even started I had to buy a new hand-mixer and that's where it all began to go wrong. I put two eggs rather than three into the sponge mix. I didn't measure my bicarbonate of soda properly, and then I found that my supply of Valrhona cocoa powder had been filched by a child, so I had to go out again and came back with inferior Green & Black's.

I didn't like the look of my mixture but I put it in the oven hopefully. For the first time in my life both cake tins suffered a total collapse when exposed to the air. It was a disaster. I had an hour and a quarter and had to start again.

I was finishing my second cake just as my rival's face appeared at the back door. I was still struggling, since I'd had to apply both my fondant filling and my top icing while the cake was warm. I put it into the fridge and played for time, hoping everything might vaguely set.

The judging panel was a serious-minded group of fifteen-year-olds from my sons' school. They were given scorecards inviting marks out of ten for appearance, taste and texture. They were to judge not two but three cakes, as my daughter and her friend had conjured a last-minute roulade filled with whipped cream and decorated with marzipan mice, one of which had a knife through its head and lay in a pool of strawberry jam.

The contestants sat outside in the sunshine as the eight judges went in one by one on a strict non-conferring basis. Being scholarly they also insisted on writing comments about each cake. The best comment on mine was 'Looks like a cake – good and tasty,' and the weakest was 'My least favourite, bad texture, didn't like it too much.'

My rival scooped 'Lovely filling, firm but still moist. Lovely cream topping.' Unfortunately she will also have to live with 'Looks like a cow pat but is mmm.'

My daughter and her friend's roulade was described as 'Well nice and chocolaty mmmmmm. Cream nice too,' whilst also being assessed 'Didn't like white bit (cream?) – didn't go with the rest. Beheaded mouse was just plain wrong.'

After an hour of careful scrutiny and independent verification of the scores we were announced in reverse order. I had 198/240, my friend's wife had 202.5/240, but our daughters had won the day with 208/240.

Many photographs were taken and cheeks were kissed in genuine congratulation. I felt absolutely shattered, and then had to endure having the piss taken out of me when my children found my first abortive effort in a cake tin and put it out for all to mock.

But where there is cake there is magic. When we travel to the desert we marvel at the traditional hospitality of the Bedouin, but we forget that we are not bad hosts ourselves. One of my son's friends had been given a lift by her parents and as the cake contest started up we made them a cup of tea and they stayed for the entire thing. Indeed, her father became chief returning officer of the voting process and lent it an aura of immense gravitas.

Later that afternoon he prepared to leave.

'I just wanted to say that this has been one of the most unusual afternoons of my life.'

English to the end I pooh-poohed him.

'No, really, a remarkable thing to do,' he insisted. 'I will remember it fondly as long as I live.'

He shook my hand again and left the house. I turned back to the kitchen to see a dozen children finishing the competing cakes so fast I didn't get a single mouthful.

I carried my mug of tea outside into the garden and had a few quiet words on my own with Daisy the rabbit. I looked back towards our house to this casually joyful scene, and realised that it had all been composed in the name of cake.

Although only Daisy could have seen it, I had to move quite fast to wipe what felt like a raindrop away from my cheek.

Next year we're doing Victoria sponge.

The Cherry on the Cake

Ever since my dad attempted to make wine in our garage with a mixture from Boots, a demijohn and length of hosepipe I have been opposed to men making at home that which is best bought from a shop. But my specialist area is another matter. Like a reluctant believer repeatedly exposed to miracles, I realise that a man who denies the joy of making his own pudding at home, merely to affirm his reverence for Mr Kipling, is arresting his own development.

BATTENBERG CAKE

I used to consider that making your own Battenberg was futile. It was begotten not created. However, the satisfaction of creating the pink and yellow check effect within a marzipan blanket ranks high in the amateur cakemakers' repertoire. I have adapted this recipe from a number of others. Only the original chef at the Victorian wedding knows the real truth, but this will bring you close.

Preparation: 90 mins
Cooking: 40 mins
Oven: 190°C (375°F, Gas Mark 5)
Serves: 6

INGREDIENTS
225g butter
225g caster sugar
4 eggs
225g self-raising flour
60ml (4 tbsp) milk
½ tsp vanilla essence
red food colouring
45ml (3 tbsp) apricot jam
caster sugar
225g almond paste

METHOD
1. Preheat the oven to 190°C (375°F, Gas Mark 5). Prepare a 9 × 6 inch Swiss Roll cake tin by cutting a piece of heavy-duty aluminium foil the exact length of the base, but 5cm (2 inches) wider than the width. Fold a 2.5cm (1 inch) pleat down the centre of the foil so it lies flat on the bottom of the tin, with the pleat standing up to form a wall which divides the tin in half lengthways. Brush the foil lightly with oil.

2. In a large bowl, beat together the butter and sugar until pale and fluffy. Beat in the eggs one at a time, adding 1 tbsp of the flour with each egg. Then add the remaining flour and gradually fold into creamed mixture with a metal spoon, adding the milk and vanilla essence and mix to a consistency that will easily drop off the spoon.

3. Pour half the mixture into one side of the prepared tin. Add a few drops of food colouring to the remaining mixture and beat until colour is even. Pour the coloured mixture into other half of the tin. Bake in the centre of the oven for 40 minutes, or until risen and firm to the touch. Turn out of tin and leave to cool on wire rack.

4. Once cooled, trim the two halves to equal size and cut each in half lengthways. Warm the jam in a small pan. Spread one strip of plain cake with jam and place one pink strip on top. Repeat process with the pink strip as the base. Stick two halves of cake together with jam. This should create a checkered effect.

5. Sprinkle icing sugar over a working surface and roll out the almond paste to a rectangle large enough to wrap around the cake. Spread the remaining jam over the outside of the cake, then wrap in the almond paste, trimming the edges to leave the ends of the cake uncovered, and seal the join well. Crimp the top edges to make a border and lightly score top with diamond pattern using a sharp knife. Let the cake stand covered for a couple of hours (preferably overnight). Serve cut into slices.

LAKE MICHIGAN CHOCOLATE FUDGE CAKE

My chocolate fudge cake recipe is American in origin. If cholesterol is an issue you'd better go back to the rice biscuits, because the twin keys to success are top-notch cocoa mulched into a fridgeload of butter. The beauty is that you get a better cake each time you bake it. There is much acquired skill in precision whisking, in taking it from the oven before it has overcooked and, in my experience, of being careful how you turn it out on to a wire rack without breaking it in two. You may know the excellent Pret A Manger version. This is better.

Preparation: 1 hour
Cooking: 20–30 mins
Chill: 4 hours or overnight
Oven: 180°C (350°F, Gas Mark 4)
Serves: 8–10

INGREDIENTS
For the cake
    85g cocoa
    354ml (12fl oz) boiling water
    3 eggs size 2
    6g vanilla extract
    300g self-raising flour, sifted
    434g light brown sugar
    7g (1¼ tsp) bicarbonate of soda
    5g (¾ tsp) salt
    227g unsalted butter, softened

For the filling
    454g milk chocolate
    227g dark chocolate
    340g unsalted butter, softened

For the icing
    75g icing sugar
    25g cocoa
    Warm water to mix

METHOD
1. Pre-heat oven and grease two 8 inch round cake tins.

2. Whisk cocoa and boiling water until smooth. Cool to room temperature.

3. In another bowl, lightly combine the eggs, ¼ of the cocoa mixture and the vanilla extract.

4. In a large bowl, combine the flour, sugar, bicarbonate of soda and salt with an electric mixer for 30 seconds. Add butter and remaining cocoa mixture. Mix at low speed until dry ingredients are moistened. Increase to medium speed and beat for 90 seconds.

5. Scrape down the sides. Gradually add the egg mixture in three batches, beating for 20 seconds after each addition.

6. Scrape batter into greased tins and smooth. Bake for 20–30 minutes until tester comes out clean.

7. Let cakes cool in tins on racks for 10 minutes. Loosen with metal spatula and invert on to greased wire wracks. To prevent splitting, re-invert so that tops are up, and allow to cool completely.

8. For the filling, break the chocolate into squares and put in a bowl. Microwave for 15 seconds at a time until soft, or melt over a pan of gently simmering water. Allow to cool.

9. Beat the butter, then beat in the cooled chocolate until uniform in colour. Use a hot metal spatula (dipped in boiling water, and dried) to spread the filling on to one of the cooled cakes. Then place the other cake on top.

10. To make the icing, mix the icing sugar and cocoa with a tablespoon of warm water. Mix with a fork and very gradually add enough water to form an easily spread mixture that will not drip down the cake's sides. The icing will thicken when cool. Using a hot spatula, spread the icing on the top of the cake, being careful not to allow it to go over the edge. Best kept in the fridge but served at room temperature. Can be microwaved on full power for 25 seconds per slice and served with vanilla ice cream as a superb pudding worthy of a dinner party.

AVOCADO CHOCOLATE MOUSSE

This pudding sounds like a practical joke, but with its use of maple syrup and overripe avocados it is perfect for all sorts of allergic types who suffer problems with gluten or dairy products.

It is astonishingly tasty and, if you can keep the main ingredient quiet until everyone has eaten the lot, gives the added frisson of making your dinner guests gag.

Preparation: 10 mins
Chill: 2 hrs
Serves: 4–6

INGREDIENTS
3 overripe avocados
100ml maple syrup
50g cocoa powder
50ml water
1 vanilla pod

METHOD
1. Spoon out flesh of avocados and place into a blender with the seeds you have scraped from the vanilla pod, the maple syrup, cocoa and water. Blend until smooth, scraping the ingredients down from the sides of the blender a few times.

2. Spoon the mousse into glasses and chill in the fridge for at least 2 hours.

3. Serve with raspberries or strawberries and fresh or sour cream (though not for you dairy-intolerants) – and don't tell anyone it's avocados until after they have eaten it.

GOOD AND EVIL PRALINE TERRINE

My mother used to make a pudding called Charlotte Katrine for my dad's occasional business guests, an almond and chocolate fondant within sponge fingers which he teased her should have been called Stonehenge. It was always a hit and I prayed there'd be a sliver left over for me the next day. My wife's equivalent of this is an extraordinary concoction of terrines and praline, and in the unlikely event we wanted to Jerry and Margot a boss one day this would surely clinch my promotion.

Preparation: 1 hour
Cooking: 10 mins
Chill: 4 hours
Oven: 180°C (350°F or Gas Mark 4)
Serves: 8–10

Although it looks complicated, you can make the white and dark chocolate mousses at the same time.

INGREDIENTS
For the praline coating
225g blanched almonds
125g granulated sugar

For the dark chocolate mousse
175g plain chocolate
75g caster sugar
75g unsalted butter, softened
75g cocoa powder
3 egg yolks
30ml Cointreau or fresh orange juice
300ml whipping cream

For the white chocolate mousse
175g white chocolate
50g unsalted butter, softened
2 egg yolks
45ml (3tbsp) crushed praline (see step 1)
200ml whipping cream

METHOD

1. To make the praline: toast the almonds on a baking tray in the oven at 180°C (350°F, Gas Mark 4) for 10 mins or until golden brown, tossing occasionally. Put the sugar and 15ml (1tbsp) water in a heavy-based saucepan over a low heat and stir until dissolved. Add the almonds and boil until the syrup starts to turn a deep golden brown. Immediately pour on to a lightly oiled baking sheet. Leave until cold, then crush in a food mixer or put in a plastic bag, cover with a tea towel and crush with a rolling pin until it resembles breadcrumbs.

2. Grease and line a 1.1 litre (2 pint) loaf tin with cling film. Melt the plain chocolate and white chocolate separately in small heatproof bowls over pans of simmering water. Stir until smooth and allow both to cool slightly.

3. To make the dark chocolate mousse, cream the butter and half the sugar together until pale. Then sift in the cocoa and beat. In a separate bowl beat together the egg yolks and remaining sugar until pale. Stir in the Cointreau or fresh orange juice. In another bowl whip the cream until stiff. Beat the melted plain chocolate into the butter mixture. Quickly stir this into the yolk mixture. Carefully fold in the whipped cream.

4. For the white chocolate mousse, beat the butter into the melted white chocolate. Stir in the egg yolks and 45ml (3tbsp) crushed praline. Whip the cream until it begins to hold its shape and fold into the white chocolate mixture.

5. To create a marbled effect, alternately place large spoonfuls of the two mousses into the prepared tin until all the mixtures are in the tin. Cover and chill overnight or for at least 4 hours.

6. Turn out the mousse on to a flat serving dish and remove the cling film. Press the praline over the top and sides until the whole mousse is covered. Chill for a further 30 minutes. Serve in slices with seasonal berries to decorate.

CUSTARD TART

With the threat from global warming and the war on terror it is important that every man learns how to make his own custard tart in the event of Marks & Spencer having to close for a few days. In America custard is a drink. The French call it *Crème Anglaise*. Tesco brand it as Egg Custard Tart, which makes it sound too savoury.

A proper British custard tart is the perfect vehicle for grated nutmeg, and the justification for all those emissions from the back of a fridge.

Preparation: 15 mins
Cooking: 15 mins + 40 mins
Chill: 15 mins
Oven: 200°C (400°F, Gas mark 6)
Serves: 6

INGREDIENTS
100g short crust pastry
1 egg white lightly beaten (keep yolk for custard)

For the Custard
3 eggs
1 egg yolk
2 tbsp vanilla sugar
300ml single cream
150ml milk
Pinch ground mace
Pinch ground nutmeg

METHOD

1. Pre-heat oven. Roll out the pastry to line a deep flan or quiche case (20cm diameter) and bake blind for 15 minutes. Brush the cooked pastry case with the lightly beaten egg white and bake for a further 5 minutes. This will seal the pastry case and prevent the custard making it soggy. Remove from the oven and turn the temperature down to 160°C (315°F, Gas Mark 3).

2. Whisk the eggs, egg yolk and sugar. Put the cream, milk, mace and egg mixture into a saucepan and warm but do not allow to boil, stirring gently.

3. Avoid spilling the mixture by putting the pie dish on the oven rack, then sieve the mixture into the pastry case. Sprinkle with grated nutmeg and carefully push the rack back into place.

4. Bake for 35–45 minutes. When done the custard should look solid but with a wobbly centre.

5. Remove from oven and allow to cool slightly before refrigerating for an hour. Serve cold.

APPLE STRUDEL

Finally, the dish which plays most profoundly in both the middle European Heimat and the flatlands of Nebraska, the pudding of emigration and of home. Impossible to get right first time, but the most likely of all recipes to allow a little personal finessing as you learn to make a whirlpool out of pastry and apple. If either a man or woman makes this for you on a date, marry them.

Cooking: 30–40 mins
Cooling: 30 mins
Oven: 190°C (375°F, Gas Mark 5)
Serves: 8

INGREDIENTS

For the pastry

225g strong white flour
2.5ml (½ level tsp) salt
1 egg, lightly beaten
30ml (2tbsp) oil
1.25ml (¼ level tsp) lemon juice
90ml (6tbsp) lukewarm water
25g butter

For the filling

450g apples (Golden Delicious or Pippin lose their shape less than others) peeled, cored and thinly sliced
75g raisins or sultanas (depending on preference)
50g coarsely chopped walnuts
65g caster sugar
2.5ml (½ level tsp) ground cinnamon
1.25ml (¼ level tsp) ground nutmeg
100g dry white breadcrumbs
50g butter

All the time for this recipe goes in the pastry making. The thinner the pastry the better the strudel and practice makes perfect. If you can see the pattern of the material on the tea towel through the dough it is probably thin enough.

METHOD

1. To make the pastry, mix flour and salt in a large bowl. Make a well in the centre and pour in egg, oil and lemon juice. Gradually add water to make a soft sticky dough. Knead the dough in the bowl until it leaves the sides clean. Turn the dough on to an unfloured surface and knead for a further 15 minutes, occasionally picking it up and throwing it back down on the surface. Return to bowl, lightly brush top with oil. Wrap tightly in cling film and let stand for 30–90 minutes (the longer the better).

2. For the filling mix together the apples, raisins, walnuts, sugar, cinnamon and nutmeg, along with half the breadcrumbs. Grease a 38 × 25cm (15 × 11inch) Swiss Roll tin.

3. Put the dough on a large, clean, floured tea towel and roll out to a very thin rectangle, lifting occasionally to prevent sticking. Brush with butter. Stretch the dough gently by putting your hands palm down between the tea towel and the dough, then gently working it with the backs of your hands from the centre to the edge. Continue to lift and stretch the dough until paper thin (slight tears do not matter) and measures at least 50 × 30cm (20 × 16ins). Trim straight with scissors.

4. Brush the pastry with half the melted butter and sprinkle with remaining breadcrumbs. Spoon on the apple mixture leaving a 7.5cm (3in) border on three sides, placing the filling to the edge of one long side.

5. Fold over the borders and use the tea towel to help you roll up lengthways from the filled edge. Lift and place it seam down in the tin. Brush with remaining butter and bake at 190°C (375°F, Gas Mark 5) for 35–40 minutes until golden.

6. Cool in the tin for 30 minutes and serve in slices with cream, ice cream or custard.

7. Await proposal.

Acknowledgments

I'd like to thank my publisher and first-class editor, Jocasta Brownlee, for her faith and remarkable skill with the blue pencil, together with the great team at Hodder for bringing this book to the shelves; my agent at PFD, Simon Trewin, for his trusty swordsmanship in the ever-changing world of publishing; the judicious young people of Colyton Grammar School for assessing our baking skills; my children, whose own sweetie day remains once a week despite their father 'fessin' up; and especially my wife, Lydia, for her endless encouragement and love, and her occasional companionship in cake.